T0313366

Ultra-Low Power FM-UWB
Transceivers for IoT

RIVER PUBLISHERS SERIES IN CIRCUITS AND SYSTEMS

Series Editors:

MASSIMO ALIOTO
National University of Singapore
Singapore

KOFI MAKINWA
Delft University of Technology
The Netherlands

DENNIS SYLVESTER
University of Michigan
USA

Indexing: All books published in this series are submitted to the Web of Science Book Citation Index (BkCI), to SCOPUS, to CrossRef and to Google Scholar for evaluation and indexing.

The "River Publishers Series in Circuits and Systems" is a series of comprehensive academic and professional books which focus on theory and applications of Circuit and Systems. This includes analog and digital integrated circuits, memory technologies, system-on-chip and processor design. The series also includes books on electronic design automation and design methodology, as well as computer aided design tools.

Books published in the series include research monographs, edited volumes, handbooks and textbooks. The books provide professionals, researchers, educators, and advanced students in the field with an invaluable insight into the latest research and developments.

Topics covered in the series include, but are by no means restricted to the following:

- Analog Integrated Circuits
- Digital Integrated Circuits
- Data Converters
- Processor Architecures
- System-on-Chip
- Memory Design
- Electronic Design Automation

For a list of other books in this series, visit www.riverpublishers.com

Ultra-Low Power FM-UWB
Transceivers for IoT

Vladimir Kopta

CSEM SA
Switzerland

Christian Enz

EPFL
Switzerland

Routledge
Taylor & Francis Group

LONDON AND NEW YORK

Published 2019 by River Publishers

River Publishers

Alsbjergvej 10, 9260 Gistrup, Denmark

www.riverpublishers.com

Distributed exclusively by Routledge

4 Park Square, Milton Park, Abingdon, Oxon OX14 4RN

605 Third Avenue, New York, NY 10017, USA

Ultra-Low Power FM-UWB Transceivers for IoT / by Vladimir Kopta, Christian Enz.

Routledge is an imprint of the Taylor & Francis Group, an informa business

ISBN 978-87-7022-144-3 (print)

While every effort is made to provide dependable information, the publisher, authors, and editors cannot be held responsible for any errors or omissions.

Contents

Foreword   ix

Preface   xi

List of Figures   xiii

List of Tables   xix

List of Abbreviations   xxi

1 Introduction   1
  1.1 Wireless Communication 1
  1.2 CMOS Technology and Scaling 2
  1.3 IoT and WSN 3
  1.4 Energy Sources 5
  1.5 Work Outline 6
  References 8

2 Low Power Wireless Communications   9
  2.1 Introduction 9
  2.2 Narrowband Communications 12
      2.2.1 Bluetooth Low Energy 13
      2.2.2 Wake-up Receivers 18
  2.3 Ultra-Wideband Communications 23
      2.3.1 Impulse Radio UWB 24
  2.4 Concluding Remarks 29
  References 31

3 FM-UWB as a Low-Power, Robust Modulation Scheme   **39**
  3.1   Introduction   39
  3.2   Principles of FM-UWB   40
       3.2.1   FM-UWB Modulation   40
       3.2.2   Multi-User Communication and Narrowband
               Interference   44
       3.2.3   Beyond Standard FM-UWB   51
  3.3   State-of-the-Art FM-UWB Transceivers   54
       3.3.1   FM-UWB Receivers   54
       3.3.2   FM-UWB Transmitters   58
       3.3.3   FM-UWB Against IR-UWB and Narrowband
               Receivers   60
  3.4   Summary   63
  References   64

4 The Approximate Zero IF Receiver Architecture   **67**
  4.1   Introduction   67
  4.2   The Uncertain IF Architecture   68
  4.3   The Approximate Zero IF Receiver with Quadrature
       Downconversion   70
  4.4   The Approximate Zero IF Receiver with Single-Ended
       Downconversion   78
  4.5   Receiver Sensitivity Estimation   82
  4.6   Summary   86
  References   87

5 Quadrature Approximate Zero-IF FM-UWB Receiver   **89**
  5.1   Introduction   89
  5.2   Receiver Architecture   89
  5.3   Circuit Implementation   92
       5.3.1   RF Frontend   92
       5.3.2   IF Amplifier   95
       5.3.3   LO Generation and Calibration   98
       5.3.4   FM Demodulator   103
       5.3.5   LF Amplifier and Output Buffer   104
       5.3.6   Current Reference PTAT Circuit   106
  5.4   Measurement Results   107
       5.4.1   General Receiver Measurements   107
       5.4.2   Single User Measurements   109

      5.4.3   Multi-User Measurements 116

      5.4.4   Multi-Channel Transmission
             Measurements 119

  5.5   Summary 120

  References 122

6 FM-UWB Transceiver   **125**

  6.1   Introduction 125

  6.2   Transceiver Architecture 125

  6.3   Transmitter Implementation 127

      6.3.1   Sub-Carrier Synthesis 128

      6.3.2   DCO Digital to Analog Converters 129

      6.3.3   DCO 132

      6.3.4   Preamplifier and Power Amplifier 134

  6.4   Receiver Implementation 139

      6.4.1   RF Frontend 139

      6.4.2   IF Amplifiers 141

      6.4.3   Receiver DCO 143

      6.4.4   Demodulator 144

      6.4.5   N-Path Channel Filter 147

      6.4.6   LF Amplifier and Comparator 153

      6.4.7   FSK Demodulator and Clock Recovery 155

      6.4.8   SAR FLL Calibration 159

      6.4.9   Clock Reference 160

  6.5   Measurement Results 162

      6.5.1   Transmitter Measurements 163

      6.5.2   Receiver Measurements 170

  6.6   Summary 183

  References 185

7 Conclusion   **189**

  7.1   Summary of Achievements 190

  7.2   Future of FM-UWB 192

  References 193

Index   **195**

About the Authors   **199**

Foreword

Low-power and ultra-low-power communication technology is enabling the internet of things (IoT). The technology described by Kopta and Enz in this book is energy-efficient, robust and offers the capability of license-free communication – all desirable attributes for future IoT devices. Wideband- and ultrawideband-FM (FM-UWB) radio transceivers have been pioneered by researchers affiliated with the Swiss Center for Electronics and Microtechnology (CSEM), and elsewhere in Europe, since 2002. The technology offers an elegant and simple solution to the energy and performance constraints for many IoT applications. Kopta and Enz continue the research tradition established by John Gerrits at CSEM in 2002–2003, and have brought UWB-FM even closer to its goal of commercial exploitation.

After a brief introduction outlining the constraints and motivation for transceivers integrated in silicon CMOS for wireless sensor networking applications, a survey of narrowband and wideband transceivers is presented in Chapter 2. Chapter 3 is devoted to a tutorial on FM-UWB, which gives the reader a concise overview of the principles behind the double-FM method of modulation/demodulation, and a review of the transceiver implementations reported in the recent literature based on FM-UWB schemes. A key advantage intrinsic to FM-UWB is network scalability. Multiple data sources can share the same RF band easily using separate FSK-modulated subcarriers. This multi-user concept was proposed by Gerrits early in his development of the FM-UWB concept, and the authors devote much of this book to their development of an experimental, low-power FM-UWB transceiver that supports multi-user scenarios.

Many of the radio technologies described in the book will be familiar to experienced CMOS practitioners. However, the authors have also provided sufficient details for the novice to easily follow their hardware demonstrator descriptions. The first designs outlined in Chapter 4 use direct conversion to baseband (i.e., zero-IF architecture) in the receiver. Rather than relying on a fixed intermediate frequency (IF), the concept of a sliding or *uncertain* IF is explained. The authors then propose an *approximate* IF receiver that leverages the uncertain-IF concept to conserve power. Both single-ended and quadrature downconversion schemes are described, and system-level simulations are presented which estimate the expected performance of the two receivers. Chapter 5 describes implementation of the quadrature approximate-zero-IF receiver concept. Circuit blocks comprising the receiver are detailed and key simulation results are presented and compared with measurements. Performance is characterized with narrowband and wideband interferers present, allowing the unique features of the FM-UWB approach to be highlighted.

While simple in concept, implementation of a practical FM-UWB transceiver requires attention to many details, and Kopta does not disappoint the reader when describing prototype implementations in 65-nm bulk CMOS in Chapter 6. Sub-carrier synthesis, the digitally-controlled carrier oscillator, and antenna amplifiers are detailed for the transmitter, including calibration schemes used to ensure robustness of the final prototypes. On the receive side, each of the circuits blocks in the receive chain are presented in depth, including an N-path channel selection filter. The emphasis in Kopta's design is on robustness to narrowband interferers, in particular interference from the 2.4-GHz ISM band. Tolerance to frequency offsets is also considered. The final prototype is able to tolerate clock offsets large enough to obviate the need for a reference oscillator, making it the first FM-UWB transceiver that can be implemented without an external quartz crystal.

In summary, readers of this book will find a complete description of the current state-of-the-art in FM-UWB technology, including detailed circuit descriptions and convincing proof-of-concept verifications of CMOS prototypes within its covers.

John R. Long
Waterloo, Ontario
November 11, 2019

Preface

This book is a result of more than four years of research work on a Ph.D. thesis at Swiss Federal Institute of Technology (EPFL) and Swiss Center for Electronics and Microtechnology (CSEM). The main motivation comes from the *WiseSkin* project, that had as a goal the integration of a sensory "skin", intended for use in prosthetic devices. Such skin would allow persons that have lost a limb to regain a natural sense of touch and perceive the artificial limb as part of their body. Tactile capability of the skin was provided by the means of a network of connected, highly miniaturized, sensor nodes, able to detect pressure and communicate data. FM-UWB imposed itself as an approach that suited all of the system needs and was quickly adopted for our solution. Beyond the scope of the *WiseSkin* project, the FM-UWB is considered here in a broader context of wireless sensor networks and IoT, topics still gaining on popularity today.

The aim of the book is to provide in depth coverage of FM-UWB as an efficient modulation scheme in the context of low-power, short range communications. It showcases FM-UWB as an alternative to commonly used narrowband radios, such as Bluetooth or ZigBee, and attempts to emphasize its potential in the IoT application space. The book also covers a design of a fully integrated FM-UWB transceiver, from high-level considerations and system specifications to transistor level design. Some of the basic concepts in circuit design are omitted in this book in order to focus on the topic of interest. The assumption is that the reader has a good foundation in analog and RF IC design, and that he is already familiar with fundamentals of communication theory. The book is intended for graduate students and academic staff engaged in electrical and electronic engineering, as well as more experienced engineers looking to expand their knowledge of low power transceivers.

The authors would like to acknowledge the members of the staff, present and past, of the Integrated and Wireless Systems Division at CSEM for their valuable contribution to this research work. It was through collaboration and many insightful discussions that the authors could benefit from their knowledge and immense experience in RF circuit design. Their helpful advice, both theoretical and practical, has proven to be crucial for the success of this work, hence special thanks go to: David Barras, David Ruffieux, Franz Pengg, Erwan Le Roux, Alexandre Vouilloz, Nicola Scolari, Nicolas Raemy, Pascal Persechini, John Farserotu, Ricardo Caseiro and Pierre-Alain Beuchat of CSEM.

The authors would also like to thank the staff at River Publishers, particularly Junko Nakajima, for the support in making this book.

Vladimir Kopta
Neuchatel, Switzerland
November 2019

List of Figures

Figure 1.1   The *WiseSkin* concept (a) and a prototype sensor
node (b)   4

Figure 2.1   Power consumption and range of different wireless
technologies   10

Figure 2.2   Typical BLE transmitter block diagram   13

Figure 2.3   Low power BLE receiver block diagrams, low IF
(a) and sliding IF (b) receiver   15

Figure 2.4   Typical WU receiver block diagram   19

Figure 2.5   Ultra wideband communication schemes, IR-UWB
(a) FM-UWB (b)   24

Figure 2.6   IR-UWB transmitter block diagram   26

Figure 2.7   IR-UWB OOK/PPM receiver block diagram   26

Figure 2.8   Low power receivers, data rate vs. power
consumption   29

Figure 2.9   Low power receivers, sensitivity vs. efficiency . . .   30

Figure 3.1   Principle of FM-UWB signal modulation   40

Figure 3.2   Wideband FM demodulator   41

Figure 3.3   Comparison of standard orthogonal FSK and
FM-UWB modulation   43

Figure 3.4   FM-UWB multi-user communication   46

Figure 3.5   FSK sub-channel frequency allocation and limits
due to distortion   49

Figure 3.6   ACPR for filtered and non-filtered FSK signal, as a
function of channel separation (100 kb/s data rate,
modulation index 1)   50

Figure 3.7   FM-UWB multi-channel broadcast   52

Figure 3.8   Example of transmission on two channels, time
domain sub-carrier signal (a) and transmitted signal
spectrum (b)   53

Figure 3.9   FM-UWB receiver architectures reported in the
literature   55

Figure 3.10   Frequency-to-amplitude conversion characteristic of reported FM demodulators   56

Figure 3.11   Comparison of FM-UWB receivers and other low power receivers from Chapter 2, data-rate against power consumption   62

Figure 3.12   Comparison of FM-UWB receivers and other low power receivers from Chapter 2, efficiency against sensitivity   62

Figure 3.13   FM-UWB transmitters and receivers, evolution of power consumption. Type of demodulator used in each receiver is indicated on the graph   64

Figure 4.1   Principle of operation of the uncertain IF receiver .   69

Figure 4.2   Block diagram of approximate zero IF receiver with IQ downconversion   71

Figure 4.3   Principle of operation of approximate zero IF receiver with IQ downconversion   73

Figure 4.4   Normalized fundamental C_1 and second harmonic amplitude C_2 at the output of the demodulator vs. the offset frequency. First harmonic is proportional to conversion gain. Four curves are plotted for four different values of the demodulator bandwidth (or equivalently different values of the delay τ)   75

Figure 4.5   Coefficient C_{MU} (a) and correction factor $|C_1|^2/C_{MU}$ for SIR (b) as functions of the frequency offset. Four curves are correspond to three different values of the demodulator bandwidth (or equivalently values of the delay τ)   78

Figure 4.6   Block diagram of approximate zero IF receiver with single-ended downconversion   79

Figure 4.7   Principle of operation of the approximate zero IF receiver with single-ended downconversion   81

Figure 4.8   Normalized fundamental C_1 and second harmonic amplitude C_2 at the output of the demodulator . . .   82

Figure 4.9   Simulated and calculated BER curves for the approximate zero IF receiver   83

Figure 4.10   Simulated and calculated BER curves with and without frequency offset for the approximate zero-IF receiver with quadrature downconversion (a) and single-ended downconversion (b)   85

Figure 5.1   Receiver block diagram   91

Figure 5.2   Schematic of the LNA/Mixer   93

Figure 5.3   Schematic of the IF amplifier, and the equivalent small-signal schematic of half circuit   95

Figure 5.4   Simulated conversion gain and noise figure of the RF and IF stages   97

Figure 5.5   Simplified schematic of the quadrature DCO   99

Figure 5.6   Simulated frequency and current consumption of the DCO 100

Figure 5.7   Schematic of the frequency divider 101

Figure 5.8   Schematic of the buffer between the DCO and the frequency divider 101

Figure 5.9   Frequency divider, waveforms at different points . . 102

Figure 5.10   Schematic of the wideband FM demodulator 103

Figure 5.11   Wideband FM demodulator, input and output waveforms 105

Figure 5.12   Schematic of the output buffer 106

Figure 5.13   Schematic of the PTAT current reference 107

Figure 5.14   Die photograph 107

Figure 5.15   Measured S_{11} parameter for different values of input capacitance 108

Figure 5.16   Measured frequency and current consumption of the DCO 109

Figure 5.17   Measurement setup 110

Figure 5.18   Measured BER curves for different carrier offset . . 112

Figure 5.19   Measured demodulator output waveform for different carrier frequency offsets 113

Figure 5.20   BER curves for different data rates 113

Figure 5.21   BER curves for different modulation order 114

Figure 5.22   Spectrum of the demodulated sub-carrier signal . . 115

Figure 5.23   Sensitivity as a function of in-band interferer power 116

Figure 5.24   BER curves for 2 FM-UWB users and varying input level between the two users 117

Figure 5.25   BER curves for different number of FM-UWB users 118

Figure 5.26   Spectrum of the demodulated sub-carrier signal, in different multi-user scenarios 118

Figure 5.27   BER curves for different number of broadcast
sub-channels   119

Figure 5.28   Spectrum of the transmitted signal, for the standard
FM-UWB and MC FM-UWB   120

Figure 5.29   Demodulated signal spectrum, with and without
spacing between adjacent sub-channels   120

Figure 6.1   Top-level block diagram of implemented
transceiver.   126

Figure 6.2   Block diagram of the implemented transmitter. . . .   128

Figure 6.3   Digital sub-carrier synthesizer.   129

Figure 6.4   Static DCO current DAC.   130

Figure 6.5   Dynamic DCO current steering DAC.   131

Figure 6.6   Dynamic DAC test output buffer.   131

Figure 6.7   Transmitter DCO with buffers.   132

Figure 6.8   Schematic of the frequency divider buffer.   133

Figure 6.9   Simulated DCO frequency, current consumption
and output voltage amplitude.   133

Figure 6.10   Preamplifier and power amplifier schematic.   136

Figure 6.11   Simulated S_{11} parameter at the RF IO.   137

Figure 6.12   Simulated power amplifier output power (a),
consumption (b) and efficiency (c) including the
preamplifier.   138

Figure 6.13   MU receiver LNA/mixer schematic.   140

Figure 6.14   LP receiver LNA/mixer schematic.   140

Figure 6.15   IFA schematic of MU and LP receiver.   141

Figure 6.16   Simulated characteristics of the LP and MU Rx
frontend.   142

Figure 6.17   LP receiver DCO schematic.   143

Figure 6.18   LP receiver DCO simulated frequency, current
consumption and output voltage.   145

Figure 6.19   MU receiver demodulator schematic.   145

Figure 6.20   LP receiver demodulator schematic.   146

Figure 6.21   LP receiver demodulator input and output
waveforms.   147

Figure 6.22   Band-pass N-path filter schematic.   149

Figure 6.23   Transconductor of the N-path filter.   151

Figure 6.24   Non-overlapping clock phases used to drive
switches and the differential switch-capacitor
array.   151

Figure 6.25   Non-overlapping clock generator.   152
Figure 6.26   Transfer function of the N-path filter.   152
Figure 6.27   Schematic of the second order cell of the LP filter
 and half circuit small signal schematic.   153
Figure 6.28   Simulated frequency characteristic of the MU and
 LP receiver LFA.   154
Figure 6.29   Comparator schematic.   155
Figure 6.30   Block diagram of the FSK demodulator and clock
 recovery circuit.   156
Figure 6.31   Simulated signals of the FSK demodulator and
 clock recovery circuit.   157
Figure 6.32   SAR FLL block diagram.   160
Figure 6.33   Example measured SAR FLL calibration cycle. . .   161
Figure 6.34   Principle of clock generation and distribution. . . .   162
Figure 6.35   SEM die photograph of the transceiver.   163
Figure 6.36   Measured sub-carrier DAC output (a) and measured
 frequency deviation of the transmitted signal (b). .   164
Figure 6.37   Frequency and power consumption of the transmit
 DCO.   165
Figure 6.38   Phase noise of the transmit DCO at 4 GHz.   166
Figure 6.39   Measured power amplifier output power (a),
 consumption (b) and efficiency (c) including the
 preamplifier.   167
Figure 6.40   Transmitted FM-UWB signal spectrum.   168
Figure 6.41   Transmit power vs. transmitter power
 consumption.   169
Figure 6.42   Measured S_{11} parameter in transmit and receive
 mode.   170
Figure 6.43   Measured frequency and power consumption of the
 MU Rx DCO.   171
Figure 6.44   Measured frequency and power consumption of the
 LP Rx DCO.   172
Figure 6.45   N-path filter measured characteristic for center
 frequency of 1.25 MHz.   172
Figure 6.46   Demodulated signal spectrum before and after
 N-path filter.   173
Figure 6.47   Test setup used for transceiver characterization. . .   175
Figure 6.48   Single user BER of the MU Rx with internal and
 external demodulator at 100 kb/s.   176

Figure 6.49 Single user BER of the LP Rx with internal and external demodulator at 100 kb/s. 177

Figure 6.50 Measurement setup and comparison of transmit and received bits. 178

Figure 6.51 Sensitivity degradation due to the presence of an in-band interferer. 178

Figure 6.52 Sensitivity degradation due to the presence of an out of band interferer at 2.4 GHz. 179

Figure 6.53 BER for a fixed input signal level with varying reference clock frequency. 180

Figure 6.54 Measured BER curves for multiple FM-UWB users of same power level, demodulated with external (a) and internal (b) demodulator. 181

Figure 6.55 Measured BER curves for two FM-UWB users of different power levels, demodulated with external (a) and internal (b) demodulator. 182

Figure 7.1 Power consumption evolution of implemented FM-UWB transmitters and receivers 190

Figure 7.2 Efficiency vs. sensitivity of implemented FM-UWB receivers 191

List of Tables

Table 2.1	Performance comparison of BLE receivers	16
Table 2.2	Performance comparison of WU receivers	21
Table 2.3	Performance comparison of IR-UWB receivers	28
Table 3.1	Performance summary of state-of-the-art FM-UWB receivers	57
Table 3.2	Performance summary of state-of-the-art FM-UWB transmitters	61
Table 5.1	Power consumption breakdown	109
Table 5.2	Comparison with the state-of-the-art receivers	121
Table 6.1	Transmitter power consumption breakdown	170
Table 6.2	MU receiver power consumption breakdown	174
Table 6.3	LP receiver power consumption breakdown	174
Table 6.4	Comparison with the state-of-the-art transceivers	184

List of Abbreviations

ACLR	Adjacent Channel Leakage Ratio
ACPR	Adjacent Channel Power Ratio
ADC	Analog to Digital Converter
AM	Amplitude Modulation
AWG	Arbitrary Waveform Generator
AWGN	Additive White Gaussian Noise
AZ-IF	Approximate Zero Intermediary Frequency
BAN	Body Area Network
BAW	Bulk Acoustic Wave
BB	Baseband
BER	Bit Error Rate
BLE	Bluetooth Low Energy
BT	Bluetooth
C-UWB	Chirp Ultra Wideband
CH	Cherry-Hooper
Clk	Clock
CDMA	Code Division Multiple Access
CMOS	Complementary Metal Oxide Semiconductor
CP	Continuous Phase
DAC	Digital to Analog Converter
DBPF	Dual Band-Pass Filter
DCO	Digitally Controlled Oscillator
DDS	Direct Digital Synthesis
DL	Delay Line
DPSK	Differential Phase Shift Keying
DQPSK	Differential Quadrature Phase Shift Keying
ED	Envelope Detector
EDR	Enhanced Data Rate
FCC	Federal Communications Commission
FDMA	Frequency Division Multiple Access
FH	Frequency Hopping

FLL	Frequency Locked Loop
FM	Frequency Modulation
FPGA	Field Programmable Gate Array
FSK	Frequency Shift Keying
GFSK	Gaussian Minimum Shift Keying
HD	High Density
I/Q	In-phase/Quadrature
IEEE	Institute of Electrical and Electronic Engineers
IF	Intermediary Frequency
IFA	Intermediary Frequency Amplifier
IO	Input-Output
IoT	Internet of Things
IR	Impulse Radio
ISM	Industrial Scientific Medical
LAN	Local Area Network
LF	Low Frequency
LNA	Low Noise Amplifier
LO	Local Oscillator
LP	Low Power
MEMS	Micro Electro-Mechanical System
MFC	Microbial Fuel Cell
MOS	Metal-Oxide Semiconductor
MPP	Maximum Power Point
MSO	Mixed Signal Oscilloscope
MU	Multi-User
NB	Narrowband
NF	Noise Figure
NFC	Near Field Communication
OFDM	Orthogonal Frequency Division Multiplex
OOK	On-Off Keying
PA	Power Amplifier
PCB	Printed Circuit Board
PDF	Probability Density Function
PHY	Physical Layer
PLL	Phase Locked Loop
PPA	Preamplifier
PPM	Pulse Position modulation
ppm	parts per million
PRR	Pulse Repetition Ratio

PSD	Power Spectral Density
PSK	Phase Shift keying
PTAT	Proportional to Absolute Temperature
PVC	Photovoltaic Cell
PVT	Process Voltage Temperature
QPSK	Quadrature Phase Shift Keying
RA-OOK	Random Alternate On-Off Keying
RF	Radio Frequency
RFID	Radio Frequency Identification
S-OOK	Synchronous On-Off Keying
SAR	Successive Approximation Register
SAW	Surface Acoustic Wave
SC	Sub-Carrier
SIF	Sliding Intermediary Frequency
SIR	Signal to Interference Ratio
SMA	Sub-Miniature version-A
SNIR	Signal to Noise and Interference Ratio
SNR	Signal to Noise Ratio
SPI	Serial Peripheral Interface
TDMA	Time Division Multiple Access
TEG	Thermoelectric Generator
UNB	Ultra Narrowband
UWB	Ultra-Wideband
VCO	Voltage Controlled Oscillator
WBAN	Wireless Body Area Network
WLAN	Wireless Local Area Network
WSN	Wireless Sensor Network
WU	Wake-Up (Receiver)

1

Introduction

1.1 Wireless Communication

The idea of using electromagnetic waves to send information over the air, without wires, dates back to the end of the 19th century. Although many scientists at that time worked on similar devices, Marconi was the first to demonstrate a working "wireless telegraph". First radio systems were large, relatively simple and only used in a handful of applications such as navigation and keeping contact with ships and, later on, planes. At first, they could only be used to transmit dots and dashes of the Morse code, the human voice followed slightly later in the beginning of the 20th century.

The evolution of wireless communication followed the technological advances in the field of electronics. The invention of the vacuum tube provided means to amplify and process the received weak signal and increase the range of communication. The first radio receivers were developed by Armstrong, who is well known for the invention of the super-regenerative receiver and the super-heterodyne receiver, and also the first to propose the use of frequency modulation. The next important step in the evolution of wireless was the invention of the transistor. This allowed for a remarkable reduction in weight and size of most electronic components, improved reliability and generally fueled the rapid development of mobile communications in the second half of the 20th century.

It was the reduction in size and cost of components that ultimately enabled the penetration of mobile devices into the consumer market and led to the huge commercial success of the mobile phone. Aside from the phone, different technologies were developed to transmit data over the air. W-LAN and Bluetooth followed quickly and provided a wireless link between all kinds of devices. Today, we have reached a point where wireless connection has become almost as pervasive as electricity. The radios have become so small that they can be seamlessly integrated into almost any object.

There are two main directions of development of modern wireless devices. The first one is toward higher speed and better performance, and is driven by technologies such as 5G, aiming to provide more data and uninterrupted connectivity to users in all corners of the world. Here, the size and energy come second, what matters most is providing high data rate and low latency to all users. The second direction is towards energy, size and cost reduction, driven by the IoT (Internet of Things). The vision of IoT is to provide worldwide network access to billions of objects surrounding us, to gather information from those objects and give them the ability to interact with their environment without human intervention. To achieve this, the performance of a radio must often be sacrificed in order to conform to the needs for miniaturization and low power consumption.

1.2 CMOS Technology and Scaling

The Complementary Metal Oxide Semiconductor (CMOS) process technology is undoubtedly the dominant IC technology used today. Initially intended for use in digital circuits, the CMOS is now used in a wide variety of analog and RF applications as well. The tremendous success of CMOS is a result of exponential technology development over more than 50 years. This exponential development is described by Moore's law [1] that states that the number of transistors on a silicon chip doubles every two years. Originally an observation, it became a principle that provided a guideline for the semiconductor industry for over 50 years.

The number of transistors on a chip increased owing to the reduction of feature sizes (transistor gate length). The increase in numbers and reduction of size also led to a drastic fall of price per transistor. Not only that, but smaller transistors make logic gates faster and are more energy efficient. With every new generation more and more digital functionality could be placed on a single chip, the speed of chips increased and the price kept dropping. This in turn enabled the revolution in computing, with ever more powerful computers becoming smaller and cheaper every year. Today, we are at a point where a hand-held device possesses more computing power than super-computers the size of a room used to. CMOS also revolutionized wireless communications, providing more power for digital signal processing, enabling more complex modulation and demodulation algorithms and fast error correction.

Low prices and high level of integration eventually led to the use of CMOS in analog and RF circuits. Unfortunately, the analog circuits did not benefit as much from technology scaling. One reason is that passive components simply cannot scale, as their size is governed by the laws of electromagnetism. Furthermore, the decreasing intrinsic gain of transistors and progressively lower supply voltage make analog design more challenging in modern deep sub-micron CMOS technology nodes. One parameter that improves with smaller transistor sizes is the transit frequency, i.e. the frequency where the current gain of transistor falls to 1. As a result, CMOS is now widely used in various RF applications and is slowly paving its way in the sub-THz domain, narrowing the gap between CMOS and dedicated RF and microwave technologies.

Technology scaling is one of the key enablers toward fulfilling the dream of bringing intelligence and connectivity to objects and deploying sensor networks consisting of billions of nodes. It is the exponential development of CMOS that brought the size and cost of integrated circuits down to the needed level, but innovative circuit design, novel radio architectures and clever system optimization are still necessary to provide energy efficient wireless connectivity.

1.3 IoT and WSN

Wireless Sensor Networks (WSN) and the Internet of Things (IoT) are two concepts that have been gaining traction over the past 20 years. The idea of using a distributed network of miniaturized sensing or actuating devices appeared with the advances in micro electro-mechanical systems (MEMS), digital signal processing and wireless communication [2]. Instead of using one or several sensors, a collaborative network of miniature sensor nodes is deployed near the phenomenon. These type of networks can find applications in a variety of fields, such as military, home automation, industrial automation, environmental, agricultural, construction, health care and so on. While many different types of wireless networks existed previously, their goal was mainly to provide the desired quality of service, whereas in a WSN the protocol must primarily ensure power conservation, given the limited resources available on each node and the fact that the nodes do not have access to the power grid in most cases.

An interesting subset of WSNs are the body area networks (BAN), where a network of sensors is distributed on, or inside a human body. These type of networks can provide remote patient monitoring, allowing the doctor to track all relevant physiological information without keeping the patient hospitalized, effectively improving his quality of life. A network of sensors can also be placed on a prosthetic limb and used to provide sensory information

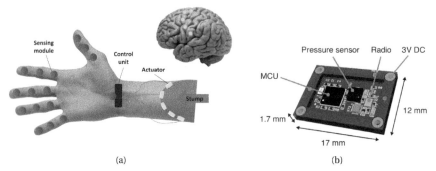

(a)   (b)

Figure 1.1   The *WiseSkin* concept (a) and a prototype sensor node (b).

to the patient, allowing him to regain the sensation of touch [3]. The tactile
information is necessary to close the feedback loop and allow the amputee
to control the limb and experience it as a part of his body. The concept,
together with the prototype of the used sensor node is shown in Figure 1.1.
As in the general case, the node itself consists of three main parts: the sensor,
the processor and the radio. The processor is the brain of the node, it stores
the information from the sensor and decides when to send this information
using the radio. In some cases local processing of the raw data can be used
to reduce the amount of data sent, reducing the overall power needed for
the radio. This is an important point as the radio is typically the dominant
consumer on a sensor node [4]. The main challenge today is to reduce the
communication power in WSNs, either by focusing on hardware and lowering
the consumption of the radio, or by focusing on software and aiming to
improve efficiency at the protocol level.

The IoT can be seen as a natural expansion of the WSN concept. In its
most basic form IoT is a vision wherein various objects are connected via
the internet. All of these objects can either extract some information about
the environment or interact with the environment using the information
provided to them. This information is stored on remote servers, where it
can be processed to extract the useful information. The IoT blends together
the internet technologies with data science and pervasive distributed sensing, with the goal of optimizing and automating industrial and common,
everyday processes. It is a commonly agreed target of the IoT to provide
approximately one thousand connected devices per person [4]. Different
challenges lay ahead before this vision can become a reality. Efficient cloud
storage and processing must accommodate the amounts of data generated by
trillions of sensor nodes. They must be able to extract knowledge from it

and provide useful services to the end user. The gateways and concentrators that provide the interface between the internet and the sensor networks must provide sufficient bandwidth to allow real time response to different stimuli. The cost and energy requirements of sensor nodes must be sufficiently low to enable deployment at such a large scale. Commercially available components are still too expensive and consume too much power to realize the IoT vision, the main obstacle currently being the wireless transceivers.

Different schemes and modulations have been used, different architectures and circuits have already been examined in order to reduce consumption of radios. This work focuses on frequency modulated ultra-wideband (FM-UWB) technology as a candidate to bring down the cost, power and size of sensor nodes, but also to enable scalability and robust communication, and bring us one step closer to realizing the IoT vision.

1.4 Energy Sources

Since most nodes in a WSN do not have access to the power grid, they must be powered in some other way. Today they are mainly powered by batteries. The exact energy needs depend on the application and dictate the size and requirements on the battery. Unfortunately, the energy capacity of batteries has not scaled as fast as the CMOS technology. Although scaling improved circuit efficiency, addition of new functions actually increased consumption of electronic devices, and batteries have simply not been able to keep up with the pace.

Different batteries are available today, providing different characteristics and using different chemistries. When designing a radio, or an entire sensor node, it is important to be aware of their characteristics in order to exploit the battery in a maximally efficient manner. The lithium based batteries are by far the mostly spread today, as they have the highest energy density of all the commercially available batteries [4]. Regardless of the type, the capacity of a battery is always proportional to its size. For a given lifetime, the node consumption determines the minimum battery size and, since the battery is often the largest component of a node, the size of the node itself. The fact that the consumption of a sensor node is proportional to its size is yet another incentive to minimize the consumption of a radio. Another issue is that the battery capacity is related to the load current, and it generally diminishes as the load current increases. High pulsed currents can significantly reduce the capacity or even damage the battery. This is highly inconvenient as the radios on a sensor node nearly always use duty cycling, meaning that they are activated to transmit a packet of data during a short period of time, after which they are put to sleep to conserve energy. As a result, the peak power of the radio has an impact on battery lifetime, making radios with lower peak current more convenient. This is why the low peak current of the FM-UWB transceiver proves to be an advantage in battery powered systems.

Batteries can be non-rechargeable and rechargeable. Rechargeable batteries can be used to extend the lifetime of a system, provided that they can be charged in some way. A very appealing mean to provide more energy to IoT devices is using the energy harvesters, especially for devices that are not easily accessible. Whenever some form of energy is available in the environment, they can be used to gather it and increase the autonomy of the node. Commercially available harvesters are thermoelectric generators (TEG), microbial fuel cells (MFC), photovoltaic cells (PVC) and piezoelectric generators. In academic work, electromagnetic harvesters, that collect energy from already present electromagnetic waves, are also described. Unfortunately, the power that can be provided by these harvesters is still too small to sustain even a small sensor node. Unsurprisingly, the amount of energy they can provide is proportional to the size of the harvester.

All of these harvesters can provide different voltage and current levels, and present a different impedance to the load. To make things more difficult, all of these quantities change with time, as the environmental conditions change. Dedicated circuits must be used to provide optimal conditions for energy extraction and track the maximum power point (MPP) of a harvester. At the same time these circuits must ensure the required voltage and current profiles in order to charge the battery in an efficient way and avoid loss of capacity. Finally, they must provide different supply voltages to different parts of a sensor node while minimizing losses. This is why the power management circuits are, alongside radios, becoming an important component of the IoT node, that is becoming increasingly more challenging to design.

1.5 Work Outline

With the previous thoughts in mind, it is the aim of this work to explore another communication scheme, the FM-UWB, within the context of WSN and IoT. The work described here was originally directed toward providing short range wireless connectivity for sensor nodes on a prosthetic limb [3], however the conclusions and results can be applied to a much broader field of applications, highlighting the potential of FM-UWB in the IoT space.

The largest part of the work focuses on the hardware implementation of an FM-UWB transceiver, emphasizing architectural and circuit techniques needed to achieve the low power consumption needed for a sensor node. Since the number of nodes in sensor networks is expected to grow in the future, the issues of scalability are also addressed. Here, the multi-user capability of FM-UWB comes into play, allowing multiple devices to share the same RF band and providing an additional degree of freedom compared to systems that exploit only TDMA (Time Division Multiple Access). The growing number of connected devices also means more devices fighting for the limited spectrum resources, and more interference in the air. The inherent rejection of narrowband interferers provided by the FM-UWB ensures a robust link even in such scenarios, allowing a transceiver implementation without a separate pre-select filter. The scaling of the CMOS technology greatly reduced the cost and size of silicon devices. The trend is therefore to integrate as many components as possible on a single die, and avoid using off-chip components that would make the node bigger and more expensive. The FM-UWB has a unique potential for full transceiver integration that gives it an advantage compared to standard narrowband transceivers. This aspect is also explored as the cost and size of devices are still a barrier to arriving at more than a thousand connected devices per person.

Chapter 2 provides a survey of low power transceivers. The most important low power techniques at the architectural level are presented and briefly explained. Performance of commercially available radios and academic implementations is shown and compared focusing on data rate, sensitivity and efficiency, but also commenting on selectivity, dynamic range and relations among these metrics. The emphasis is on architectural trade-offs of transceiver implementations as well as advantages and disadvantages of different communication schemes.

Chapter 3 explains the fundamentals of FM-UWB modulation and demodulation, highlighting the potential of this technique in the WSN and IoT context. The potential directions of evolution of FM-UWB radios are given and discussed. A survey of existing FM-UWB transceiver implementations provides a glimpse into the practical capabilities of this modulation scheme.

Chapter 4 presents the two developed receiver architectures. The "approximate zero IF" (AZ-IF) receiver architecture is derived from the "uncertain IF" architecture, previously employed to reduce consumption of the narrowband wake-up radios. The principle of operation is explained, and combined with system-level simulations in order to estimate the expected performance of the two receivers.

Chapter 5 describes the first implementation of the low power, quadrature approximate zero IF receiver. First, all the circuits are explained in detail together with the most important simulation results. Then the measurement results are reported. The receiver performance is characterized in different conditions, with narrowband and wideband interferers and using different variations of FM-UWB modulation, demonstrating in that way some of the interesting features of the FM-UWB approach.

Chapter 6 describes the implementation of the fully integrated, low power FM-UWB transceiver. Aside from the quadrature AZ-IF receiver, a single-ended receiver is added, providing a mode with even lower power consumption. As in the previous case, the two receivers are characterized under different operating conditions. In this case the emphasis is on robustness to narrowband interferers (with a special focus on 2.4 GHz band), and frequency offset tolerance. The implemented receiver is proven to tolerate clock offsets large enough to make use of an external quartz reference unnecessary, making it the first FM-UWB transceiver that can truly be implemented with no external components.

Chapter 7 concludes the topic, providing a summary of achieved results and contributions, and pointing to potential research topics and development directions for future work.

References

[1] G. E. Moore et al., "Cramming more components onto integrated circuits," 1965.

[2] I. F. Akyildiz, W. Su, Y. Sankarasubramaniam, and E. Cayirci, "Wireless sensor networks: a survey," *Computer Networks*, vol. 38, no. 4, pp. 393–422, 2002.

[3] J. Farserotu, J. Baborowski, J. D. Decotignie, P. Dallemagne, C. Enz, F. Sebelius, B. Rosen, C. Antfolk, G. Lundborg, A. Björkman, T. Knieling, and P. Gulde, "Smart skin for tactile prosthetics," in *2012 6th International Symposium on Medical Information and Communication Technology (ISMICT)*, Mar. 2012, pp. 1–8.

[4] M. Alioto, *Enabling the Internet of Things: From Integrated Circuits to Integrated Systems*. Springer International Publishing, 2017.

[5] J. Gubbi, R. Buyya, S. Marusic, and M. Palaniswami, "Internet of things (IoT): A vision, architectural elements, and future directions," *Future Generation Computer Systems*, vol. 29, no. 7, pp. 1645–1660, 2013.

2

Low Power Wireless Communications

2.1 Introduction

Reducing the power consumption of radios has been a topic of research for quite a long time. In recent years, the search for energy efficient means of communication and constant effort to lower the power consumption of wireless transceivers have been primarily driven by the growing popularity of the internet of things [1]. Still today, the consumption of radios is one of the primary obstacles to making the IoT concept a reality. Since the available energy sources have a very limited capacity, especially in size-critical applications, reducing the power consumption of sensor nodes remains the only mean of increasing their autonomy.

A wide range of wireless technologies exist that provide a different set of capabilities tailored for different applications. Wireless devices typically trade power consumption, data rate and sensitivity (which can be directly translated to range), although there are other important quantities, such as dynamic range and interference rejection, and finally size and cost, that play an important role. Some of the well-known wireless technologies are presented in Figure 2.1, that gives a qualitative comparison of consumption and range. Typically, larger range requires larger power consumption. However, a third axis would be needed to show a more complete picture. This third axis would be the data rate (or alternatively bandwidth). LoRa, SigFox and NB IoT devices manage to extend the range without an increase in power consumption at the cost of data rate, which further means that only a limited amount of information can be transmitted within a given time. These technologies were intended for use in wide area networks, to gather data from sensors that only need to transmit several bits every few hours (e.g. temperature sensors). Data rates that can be found in these radios typically vary from only 10 b/s up to 50 kb/s, which is why they are also referred to as the ultra narrowband (UNB) radios.

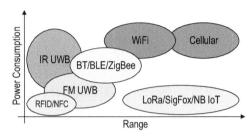

Figure 2.1   Power consumption and range of different wireless technologies.

Radio Frequency Identification (RFID) and Near Field Communication (NFC) provide the lowest power consumption. This is mainly owing to the low communication range, that is sometimes only a few centimeters, and the asymmetric link, where the burden is placed on the reader, allowing to simplify the tag. The RFID/NFC tags can consume zero power, meaning that they can be entirely powered by the reader, but their capabilities are also limited to sending only a few bytes of information, such as the ID of the tag. An RFID tag can be made very small and is often dominated by the antenna size (especially in the sub-GHz bands), whereas a reader is quite large and consumes a lot of power as it may need to transmit at watt levels.

A similar kind of asymmetry can be found in cellular networks. In this case most of the burden is placed on the base station that has access to the power grid, making its size and consumption a less important issue. This allows to reduce the size and complexity of the battery-powered, hand-held devices, while still maintaining a link over several kilometers and providing data rates on the order of 100 Mb/s. The principle is similar to the WiFi, with the burden mainly placed on the fixed device, allowing higher autonomy for the mobile devices. The difference compared to the cellular technologies is the range.

Unlike the previously mentioned standards, Bluetooth (and it's derivation Bluetooth Low Energy or BLE) and ZigBee provide a fully symmetrical link, meaning that the same transceiver is used on both sides. These radios are commonly used to provide low power wireless connectivity between two or more battery powered devices. In particular, Bluetooth has become the most widely used standard for low power connectivity in consumer applications. It provides complete wireless connectivity services, not just wireless transport, and is commonly implemented in audio streaming, data transfer and broadcasting devices. Although initially designed for small networks, consisting of up to seven peripherals connected to a master device (such as a computer, or a smart phone), Bluetooth has evolved over the years and now supports different network topologies with a much larger number of devices. It targets ranges from tens to hundreds of meters, and data rate in the order of 1 Mb/s. The fact that Bluetooth is already a dominant technology in the consumers' market means that most manufacturers want to use it in their devices regardless of whether it is really optimal for a specific application. This greatly reduces the effort and time to market that would otherwise be necessary with a custom made radio.

ZigBee, or the IEEE 802.15.4, is a similar standard with the same communication distance and lower data rates that go up to 250 kb/s. Unlike Bluetooth, it targets mainly industrial applications, and therefore the emphasis is on security, robustness and scalability, providing support for high node counts. As of 2012, enhanced ZigBee specifications also include secured connectivity to batteryless devices, powered from energy sources like motion, light or vibration.

An interesting candidate for low power communications are ultra-wideband (UWB) devices. The particularity of these devices is that the data rate is not directly linked to the bandwidth of the transmitted signals. When talking about UWB the occupied spectrum is typically much larger than required by the communication speed. Although it seems like they are wasting a very precious resource, they provide capabilities that cannot be easily achieved with classical narrowband radios. Large bandwidth relaxes constraints on the precision of the frequency synthesizer. This allows the implementation of radios without Phase Locked Lops (PLL) resulting in higher agility and making them more suitable for duty cycling. The same property also eliminates the need for a precise frequency reference, such as a quartz oscillator, allowing the reduction in size and cost of the product. UWB signals are more difficult to detect and jam, and due to their low Power Spectral Density (PSD) do not interfere with neighboring narrowband signals. Furthermore, they are capable of resisting strong interferers and operating in multi-path environments without a severe loss in performance.

In principle, the UWB radios that are in the focus of this book (targeting low to medium data rates and low power) fit between the BLE and RFID devices. They are not able to achieve the same sensitivity at the same data rate as the BLE, but they can provide a lower power consumption, size and cost. At the same time, there are no fully passive UWB solutions that can compete in the RFID space, but they provide a symmetric link, longer range and better overall performance.

Before going into the in depth analysis of the FM-UWB, this chapter will summarize and compare the key radios targeting a similar application space as the FM-UWB. The emphasis is on low power techniques, targeting low to medium data rates and a fully symmetric link. Since the network performance is almost entirely determined by the receiver performance, the rest of the chapter will be devoted to them.

2.2 Narrowband Communications

Many different narrowband techniques have been developed targeting different WSN or IoT scenarios. Today, the absolute leader in the consumers' market is the Bluetooth, which can be found in almost every portable device. ZigBee found its market in industrial automation and puts more emphasis on robustness and security. Both of these standards target distances up to 100 m at moderate data rates.

Several recently emerged standards, such as SigFox, LoRa and NB IoT are developed to cover lower data rates and provide a range of several kilometers. This is typically done by reducing the data rate. Since the probability of error depends on energy per bit E_b, that is a product of signal power and symbol duration, communication range can easily be extended by using longer symbol duration. Symbol duration is inversely proportional to the signal bandwidth, meaning that very long symbols occupy a very narrow frequency band, hence the name ultra narrowband. The commercial UNB receivers consume a similar amount of power as BLE transceivers [2] but, owing to lower data rates, achieve sensitivity better than -140 dBm (compared to BLE receivers that fall in the range between -80 dBm and -100 dBm). Combined with the output power specified above 10 dBm, LoRa radios typically provide a link budget above 150 dB.

An interesting class of receivers are the wake-up (WU) receivers. These have mainly been a topic of academic work, and haven't yet been truly adopted by the industry. Unlike the previously mentioned receivers, the WU receivers generally do not comply with any of the existing standards, therefore allowing much more freedom in optimizing different system parameters. The idea behind wake-up receivers is to constantly listen to the channel and check whether there is a signal being transmitted. Once they detect the presence of a correct sequence, they turn on the main receiver that will be used for data reception. Clearly, for the concept to work, the wake-up receivers must consume a very small amount of power. This usually comes at a price in sensitivity and speed, which is an acceptable trade-off in this case. Since WU and BLE receivers have have gained a lot of traction in recent years they will be the subject of study throughout this section.

2.2.1 Bluetooth Low Energy

Aside from dominating the market, Bluetooth and BLE transceivers are often a topic of academic work exploring new potential architectures and looking into different trade-offs between power consumption and performance. BLE is typically targeting a data rate of 1 Mb/s, although most commercial devices support higher data rates. It is intended as a standard that provides a low power connection between portable devices and sensor nodes.

The BLE uses the ISM band located between 2.4 GHz and 2.48 GHz. The used band is split into 40 channels, each 2 MHz wide. The data is transmitted at 1 Mb/s using the Gaussian Frequency Shift Keying (GFSK), meaning that a Gaussian filter is used for pulse shaping. A 2 Mb/s option is added in the Bluetooth 5 specification. An optional Enhanced Data Rate (EDR) mode provides higher data-rates by using DQPSK and 8DPSK modulations. Output power is limited to 100 mW, or 20 dBm, this limit was increased from 10 dBm in the BT 5 specification in order to extend the maximum range to above 100 m. Minimum sensitivity of BLE devices is set to -70 dBm, although most devices are providing at least -80 dBm. Frequency hopping is used to avoid collisions with other BLE transceivers or unwanted interferers.

The relatively loose BLE requirements allow a simple transmitter architecture that can be optimized for low power consumption. A typical BLE transmitter block diagram is shown in Figure 2.2. The transmitted RF signal is directly synthesized by a PLL. This approach allows to save power by avoiding mixers and quadrature signal upconversion. The modulated signal can be injected in different points of the PLL. Figure 2.2 shows the case where the VCO/DCO is directly driven by the modulated signal. A different

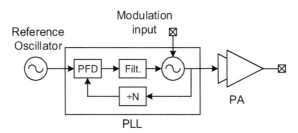

Figure 2.2 Typical BLE transmitter block diagram.

approach would be to inject the signal at the fractional frequency divider (usually controlled by a sigma-delta modulator). Injecting a signal at this point provides a good frequency control, but the approach suffers from the limited PLL bandwidth. Increasing the loop bandwidth to accommodate the needed signal bandwidth increases phase noise of the PLL. For this reason a hybrid approach has been used in some implementations, and is referred to as the two point modulation [3]. In this case signal is partially injected directly to the VCO to compensate for the limited bandwidth of the PLL. One of the advantages of the GFSK modulation, used by BLE, is that it is a constant envelope modulation. This means that the signal amplitude remains unchanged during transmission, which allows for an efficient implementation of the power amplifier. In general, power amplifiers are highly dependent on the characteristics of the transmitted signal. Non-constant envelope signals, such as $\pi/4$-DQPSK, require linear power amplifiers which are always less efficient. The advantage of non-constant envelope signals is better spectral efficiency, meaning that they occupy a smaller frequency band and leak less into adjacent channels. Efficient spectrum utilization is an important issue in high-performance systems, while it is usually sacrificed in low-power systems in order to obtain better power efficiency.

The two most commonly used receiver architectures in BLE receivers are the sliding IF and the low IF architectures shown in Figure 2.3. The low IF receiver was derived from the zero IF (or direct conversion) receiver. As the name says, such a receiver converts the input RF signal directly to zero [4]. In order to prevent a loss of information, the receiver must perform a quadrature conversion, in other words it must separate the baseband signals by their phases. The advantages of the direct conversion receiver are the absence of image (interfering signal symmetrical to the useful signal with respect to the carrier frequency), simple architecture and use of low pass filters for channel selection. However, since the baseband signal is located around zero, these receivers suffer from increased flicker noise, LO leakage, dc offset and are sensitive to I/Q mismatch. Most of these drawbacks are resolved using the low IF receiver. Centering the baseband signal between 1 MHz and 2 MHz mitigates the problems with flicker noise and dc offset. An issue with a low IF receiver is that an image now appears at negative frequencies. This image must be suppressed by a sharp complex baseband filter (usually a polyphase filter) prior to digitizing the signal. When it comes to power consumption some authors claim that this type of receiver consumes a lot of power for frequency generation, as it needs to provide quadrature signals at the RF channel frequency [5, 6]. From this point of view the sliding IF receiver has

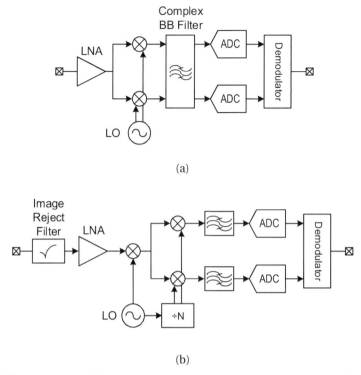

Figure 2.3   Low power BLE receiver block diagrams, low IF (a) and sliding IF (b) receiver.

some interesting properties. Although the conversion is done in two steps, only one oscillator is used. The second LO is derived from the first using a fixed frequency divider. That mans that if the input frequency changes the IF changes as well, hence the name sliding IF. If the division ratio is even, quadrature signals are available practically for free. As frequency dividers are in any case needed for the PLL, second LO generation doesn't cause additional burden in terms of power. Since the first LO frequency is lower than the input signal frequency, the consumption of the VCO and the LO buffers is somewhat reduced. For example if the fixed division ration is 2 then the LO frequency should be $f_{LO} = 2/3 f_{in}$. Additional gain comes from the fact that the VCO doesn't need to generate a quadrature signal. As with all heterodyne receivers, the image might pose problems for the sliding IF receiver as well. This issue is commonly addressed by placing an external image reject filter that suppresses all signals at the image frequency.

Table 2.1   Performance comparison of BLE receivers

Ref.	Sens. [dBm]	Cons. [mW]	Eff. [nJ/b]	Data Rate [Mb/s]	Supply [V]	Mod.
Academic Work						
[10]	−95	0.98	0.98	1	1	GFSK
[11]*	−80	0.23	2.77	0.083	0.75	FSK
[6]	−98	3.8	3.8	1	1.2	GFSK
[5]	−90	5.5	5.5	1	0.6/1.2	GFSK
[3]	−94.5	6.3	6.3	1	3	GFSK
[12]*	−57.5	0.15	1.3	0.115	0.9/1.1	OOK
Commercial Devices						
TI CC2540 [13]	−93	39	19.5	<2	2–3.6	GFSK
PSoC 63 [7]	−95	22	11	<2	1.7–3.6	GFSK
nRF51822 [14]	−93	23	11.5	<2	1.8–3.6	GFSK
nRF52832 [15]	−96	16.2	8.1	<2	1.7–3.6	GFSK
EM 9301 [16]	−83	27	13.5	<2	1.9–3.6	GFSK
DA14580 [8]	−93	11.4	5.7	<2	0.9–3.45	GFSK
TC35679 [17]	−93.5	9.9	4.95	<2	1.8–3.6	GFSK
TC35681 [18]	−95.6	15.6	7.8	<2	1.8–3.6	GFSK
RSL 10 [9]	−94	9	4.5	<2	1.1–3–3	GFSK

For all commercial devices reported sensitivity is at 1 Mb/s.
*BLE compatible WU receiver.

An overview of existing BLE receivers is given in Table 2.1. Key performance metrics are given for selected academic work, as well as for commercially available components. As seen in the table, most of commercial devices target sensitivity better than −90 dBm. The power consumption varies from 39 mW down to 9 mW in some of more recent implementations. It should also be noted that some of these implementations are parts of entire systems on chip (SoC), targeting IoT type of applications [7–9]. Most of them also come with an integrated DC-DC converter that converts the 3.3 V battery supply into voltages around 1 V, typically used by today's low power radios, and provide a low power sleep mode, making them suitable for duty cycling.

Seeing the consumption of devices reported in academic work it is clear that there is still room for improvement. Various techniques are found that enable different improvements. The receiver from [10] consumes less than 1 mW, while achieving sensitivity comparable to commercial devices. To achieve such low power the design relies heavily on current reuse. The input gain stages are stacked to maximize gain for a given bias current. Direct conversion receiver is used and the PLL provides an LO at twice the RF frequency. Such strategy is often seen in BLE radios as it reduces frequency pulling (frequency drift due to coupling to external signals) and provides an easy way to generate a quadrature LO. The design also demonstrates other useful tricks, such as placing the VCO circuitry inside the inductor to conserve area. This is possible because the magnetic field is mainly concentrated close to the conductor, and therefore the circuits in the middle have minimal effect on the inductor parameters. In this case the Q factor is degraded by 6%. In [3] a full BLE transceiver with an integrated DC-DC converter is presented. What makes this work interesting is that the LNA input and the PA output are sharing the same pad. All the passive components of the IO matching network are placed on chip, the only off-chip components required (aside from the crystal resonator) are the inductor and capacitor for the DC-DC converter. The matching network can be configured for transmit or receive mode using MOS switches. In the receive mode it also acts as a notch filter that attenuates the signal at the image frequency, hence eliminating the need for another external component.

The two receivers from [11] and [12] are not true BLE receivers in the sense that they do not conform to all the standard requirements, but they are BLE compatible receivers. This means that they can be used to decode information from a BLE compliant packet, but they are not as constrained as regular BLE receivers, allowing them to further reduce power consumption. They can be used to sample the RF channel instead of the main receiver and only wake up the main receiver when useful data is transmitted, thereby allowing global power reduction in the system. As seen in the table, they use lower data rates than those specified by the standard, allowing to reduce power while maintaining sensitivity. In order to allow for lower data rates, BLE packets on the transmitter side must be formed in a specific way (e.g. repeating the same bit several times), but such that they still conform to the standard. In work from [11], the input sequence is pre-coded such that after data whitening, that is mandatory in BLE, the output sequence gives a desired bit repetition to reduce the equivalent data rate. Following this, since the BLE packets have a limited payload length, there is a trade-off between data rate and the wake-up sequence length. The receiver from [11] uses a free running LC oscillator that is only calibrated once. Removing the PLL reduces power consumption, but also allows the receiver to start more quickly as there is no limit coming from the loop bandwidth. This approach allows to duty cycle the receiver at a bit level, further reducing the consumption, but also loosing approximately 3 dB in sensitivity. The problem with a free running LC oscillator is the precision, so the carrier offset must be tracked and corrected at IF. Given the small spacing between BLE channels the receiver could easily lock onto an undesired channel, which is the main downside of the receiver from [11]. The second BLE compatible receiver described in [12] uses quite a different approach. Instead of demodulating bits, it is detecting presence of the BLE signal on different advertising channels. The receiver is not capable of demodulating a GFSK signal, instead it is a simple envelope detector (ED) that detects the signal level at a selected BLE channel. In order to wake up the main receiver, a specific sequence of transmissions on different advertising channels must be sent. This approach allows for power reduction, owing to the demodulator simplicity, however the sensitivity is greatly degraded compared to similar receivers. Nevertheless, the work shows a very interesting way to exploit the available degrees of freedom, while providing compatibility with the BLE.

The Bluetooth standard has evolved significantly over the past two decades in order to adapt to the modern needs of a low power radio. At the same time, the hardware itself has undergone significant changes in a long lasting attempt to reduce size and power while improving performance. Still, while having a standard facilitates widespread use of BLE devices, it also imposes constraints and limits the achievable degrees of freedom. Once the constraints are gone, a large variety of different performances can be achieved. This is shown in the following section on wake-up receivers.

2.2.2 Wake-up Receivers

As the name suggests, the wake up receivers are used to listen to the communication channel and wake up the main receiver when data needs to be sent. The idea behind this is to reduce the overall system power consumption by keeping the power consuming main radio off most of the time. Clearly, the consumption of the WU receiver must be sufficiently low for such a system to be viable. For comparison, one can think of a typical BLE receiver that consumes 10 mW. If such a receiver was in a small sensor node, and if it was on all the time it would drain the battery quickly. Since in most cases the data traffic is quite low (although this depends on the application) the receiver can be kept off most of the time, and only turned on when there is a need. This is called duty cycling. The question is: how to know when to turn the receiver on? One way is to synchronize all nodes in a network. A receiver would then turn on at specific time instants to check if the data is transmitted. However, maintaining synchronization in a network often results in a significant overhead that leads to increased power consumption. The communication can also be initiated asynchronously, but in this case the transmitter needs to transmit

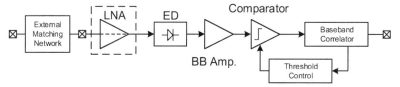

Figure 2.4   Typical WU receiver block diagram.

continuously until the receiver replies. Now the transmitter needs to consume more on average to initiate communication, but the synchronization overhead is gone. Which option is better depends on a particular case. Either way, assuming that the communication is initiated rarely, the overall power will be dominated by the receiver consumption, as it must be turned on periodically to sample the RF channel. The power consumption is directly proportional to the duty cycle (or the on time) of the receiver. If a 10 mW BLE receiver is turned on for 1 ms every second, in the idealized case the power would be reduced to 10 μW. A note should be made here that receivers cannot start immediately and that there is always an overhead associated to the start-up that would increase the average consumption. In addition, in the described scenario the worst case latency would be almost 1 s. It can easily be seen that in duty cycled systems there is always a trade-off between consumption and latency. This kind of trade-off can be broken using a wake up receiver. But, as already explained, their power consumption must be brought down to a microwatt level or below.

   To achieve the low power levels needed, the WU receivers often employ a very simple architecture and a very simple modulation scheme. In most cases, the RF signal is modulated using the on-off keying (OOK), where the presence of the RF signal corresponds to symbol 1 and the absence of it to 0. Such a signal can be easily demodulated using an envelope detector. A typical WU receiver architecture is presented in Figure 2.4. Envelope detector is followed by a correlator that compares the received sequence to the reference pattern. If the correlator output is higher than a certain value the main receiver is woken up. This threshold can be adjusted to set the ratio of false alarms and missed packets. It can be seen that the only active block operating at RF frequency is the LNA. When it comes to WU receivers the emphasis is usually on minimizing the number of blocks that operate at high frequencies, as higher frequency always mean higher consumption. In some more aggressive approaches even the LNA is omitted, allowing to go down to nanowatt consumption, but this always comes at a price in noise figure

(NF) and sensitivity. The non-linear characteristic of the envelope detector makes the output signal to noise ratio (SNR) dependent on the level of input signal. As a result the equivalent NF increases at lower signal levels, thus quickly degrading performance in the absence of an LNA. Unfortunately, this effect cannot be avoided as some kind of non-linearity is needed to separate the envelope from the RF signal. The situation can be somewhat improved if a high quality input matching network is available to boost the input voltage of the ED, but costly and relatively large off-chip components are needed. Boosting the input voltage using the on-chip passive components is impossible due to a limited Q-factor available.

The second benefit of the simple receiver architecture is the absence of the LO and the frequency synthesizer. Naturally these blocks consume a significant part of the overall power and removing them provides significant savings. This also makes the receiver more agile in the sense that start up interval is much shorter. In standard BLE receivers there are two limiting factors to increasing the start up time. The first one is the crystal oscillator. Such oscillator is necessary in all the NB radios to provide a precise reference from which all the other frequencies are derived. Because this reference needs high precision and low jitter, a high quality resonator, such as a quartz crystal, must be used. Unfortunately, the start-up time of the oscillator is inversely proportional to the Q-factor of the resonator and proportional to the oscillation frequency, resulting in a very slow start-up time. To avoid this, in some cases the reference oscillator is kept on even when the receiver switches off. If a 32 kHz clock is sufficient in the system, keeping it on all the time might cost only 100 nW, in which case the overhead might not be that severe. In most BLE receivers, however, the PLL needs a higher frequency reference that operates at megahertz frequencies and therefore consumes power in the order of microwatts. In these cases keeping the quartz oscillator always on would dominate the receiver consumption and is commonly avoided. The second limit to fast start-up is the loop bandwidth of the PLL. The lower the bandwidth of the PLL the longer it takes for the output frequency to settle, and the reception is not possible until the LO signal is stable. The settling time can be decreased by increasing the bandwidth, but this comes with increased phase noise of the PLL, which might limit the receiver performance. Finally, the simplest way to make the receiver agile is to entirely avoid LO generation. This doesn't come for free, and as a consequence, WU receivers based on an ED usually have poor selectivity and interference rejection (unless a sharp external filter is used).

Table 2.2 Performance comparison of WU receivers

Ref.	Sens. [dBm]	Cons. [μW]	Eff. [nJ/b]	Data Rate [kb/s]	Freq. [MHz]	Supply [V]	Mod.
[19]	−69	0.0045	0.015	0.3	113	0.4	OOK
[20]	−71	2.4	0.120	20	868	1	OOK
[21]	−65	10	0.100	100	2400	0.5/1*	OOK
[22]	−82	415	0.830	500	2400	1.2	IR–PPM
[23]	−71	0.0074	0.037	0.2	433	1/0.6*	OOK
[24]	−80.5	0.0061	0.183	0.033	109	0.4	OOK
[25]	−83	3	47	0.064	868	2.5	OOK
[26]	−72	52	0.500	100	2000	0.5	OOK
[27]	−41	0.098	0.980	100	915	1.2	OOK
[28]	−45.5	0.116	0.009	12.5	403	1.2/0.5*	OOK
[29]	−70	44	0.220	200	410	1	FSK
[30]	−59	0.236	0.028	8.192	2400	1/0.5*	BLE BC
[31]	−100.5	400	80	5	1900	0.9	OOK
[32]	−81	123	1.23	100	915	1	OOK
[33]	−65	2500	2.5	1000	915	0.8	OOK
[34]	−87	44.2	0.884	50	925	0.7	OOK
[35]	−75	350	0.180	2000	2400	0.65	BFSK
[36]	−97	99	9.9	10	2400	0.5	OOK
[37]	−83	227	0.227	1000	2400	0.6	OOK

*Analog/digital supply.

A comparison of recent state of the art WU receivers is given in Table 2.2. A large variety of operating frequencies, data rates and efficiencies can be observed. A large number of implementations is targeting the license free 2.4 GHz ISM band. As already stated, most of them are using OOK modulation, with several exceptions using PPM (pulse position modulation) and FSK. Power consumption varies from 2.5 mW all the way down to 4.5 nW. Different architectural and circuit techniques are used to arrive to such low power levels.

The receivers from [22, 26, 36] use a so called "uncertain IF" architecture. The name comes from the fact that the LO frequency is calibrated to be within a given range, and is not as precise as in conventional receivers, making the exact IF unknown. Low power free running oscillators are used to generate the LO. All of them are mixer first receivers that place gain stages at IF instead of RF in order to lower consumption. The IF amplifiers are wideband (100 MHz in [26]) as they must cover the whole range of possible IF frequencies. For interferer suppression, an external bulk acoustic wave (BAW) filter is used in [26], while [36] uses a series of n-path filters. A polyphase IF filter is used in [22] to provide rejection at the image frequency.

Instead of OOK, in [22] a spread spectrum technique is used. The signal is modulated using short pulses with a bandwidth of 12.5 MHz combined with PPM. The receiver can operate using a frequency reference with up to $\pm 0.5\%$ error, making it possible to remove the quartz crystal from the system, reducing both cost and size. However, the power consumption of 415 μW is higher than most other WU receivers. The implementation from [25] is a super-heterodyne receiver that uses bit-level duty cycling of all RF blocks, allowing it to reduce power consumption from 100 μW to only 3 μW. All circuits are optimized for fast start-up to minimize the overhead coming from duty cycling. One downside is the reduction of sensitivity, since the receiver is not exploiting the full energy of each symbol. The only FSK receivers are described in [29, 35]. Receiver from [35] is a super-regenerative receiver that injects the received signal into a VCO. Depending on the received frequency and the resonance frequency of the VCO the oscillations will build up faster or slower. This approach is used to distinguish between the two FSK symbols. A low IF approach is used in [29], where a low power technique for LO generation is presented. A ring oscillator that produces 9 phases is locked onto a reference oscillator using injection locking. These 9 phases are then combined to produce the LO at a frequency that is 9 times higher than the reference frequency. This approach proves to be more efficient than using an oscillator that operates directly at RF. Another super-regenerative receiver is reported in [31], here used to detect an OOK signal. This receiver achieves the highest sensitivity of all the implementations reported here. The SRR receivers achieve good sensitivity for a given power, but their consumption cannot be lowered beyond a certain point as they require an RF oscillator.

Implementations from [27, 28] were among the first that managed to reduce the power to the level of 100 nW, allowing to supply these receivers directly from an energy harvesting source. However, in these first attempts the sensitivity was quite low, in the order of -40 dBm, limiting the range and scope of use of such receivers. In [30] sensitivity was improved, while maintaining similar power consumption. Finally, receivers from [19, 23, 24] all consume below 10 nW and achieve sensitivities close to -70 dBm. One of the enablers for such low consumption is the low frequency of operation between 100 MHz and 400 MHz. At this frequencies high-Q external passive components can be used to provide large passive voltage gain before the envelope detector. Still, the sensitivity improvement mainly comes from data rates below 0.2 kb/s. In [19] the signaling speed is only 15 b/s. Assuming the wake up sequence is 15 bits long it would take 1 s to wake up the main receiver. In all of the mentioned cases low power and decent sensitivity are traded for long wake up time and increased receiver size. These WU receivers could be compared to a BLE receiver at a 0.1% duty cycle, that would be turned on for 1 ms each second. In a worst case scenario with continuous transmission, this leads to 1 s latency. Referring to the lowest power BLE receiver from Table 2.1, this receiver would consume an average power of approximately 1 μW, which is still 2 orders of magnitude above the WU receiver. In return the BLE receiver provides better sensitivity.

Different performance trade-offs can be observed in the listed WU receivers. It is important to notice that it is impossible to achieve low power, good sensitivity and high speed at the same time. WU receivers tend to sacrifice all other aspects in order to minimize power. This is likely why they haven't been widely adopted by the industry so far and remain only a subject of academic research. Various other aspects such as selectivity and interference rejection must be considered before these kind of receivers could find use in practical applications.

2.3 Ultra-Wideband Communications

The ultra wideband signals are defined as signals whose -10 dB bandwidth is either larger than 20% of the carrier frequency, or at least 500 MHz. Historically speaking UWB is one of the oldest wireless technologies – some of the first attempts to communicate over the air were done using short pulses generated by a spark gap transmitter. Today, UWB is mainly used for wireless sensing, radars and localization. The UWB radios became popular again from the communications perspective owing to several different reasons. Perhaps the most obvious one is that large bandwidth can accommodate large data rates. The UWB can be used to transmit large quantities of data over short distances in unlicensed parts of the spectrum. For security and military applications UWB is attractive because of the low power spectral density that makes the signal difficult to intercept and jam. When it comes to low power communications UWB is interesting because it offers potential for miniaturization and low power consumption, owing to the simple radio architecture, as well as robustness to interference and multipath fading.

There are different communication schemes that implement UWB, two are important with regards to wireless sensor networks and IoT. These are the frequency modulated UWB (FM-UWB) and the impulse radio UWB (IR-UWB). The two modulation schemes are illustrated in time and frequency domain in Figure 2.5. The IR-UWB uses short pulses in time domain, the shorter the pulses are the larger is the signal bandwidth. Some form of pulse

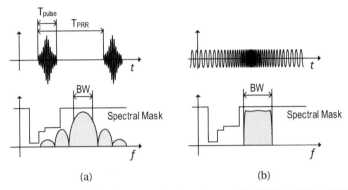

Figure 2.5 Ultra wideband communication schemes, IR-UWB (a) FM-UWB (b).

shaping is usually employed to reduce the side lobes and provide better frequency characteristics. Since the pulse is only present for a short period of time, probability of detection and interfering with other signals is low, making the IR-UWB fairly robust.

The FM-UWB can be seen as a direct frequency domain approach where a slow moving RF carrier is sweeping a large frequency range. Compared to IR-UWB, in this case a continuous signal is transmitted, providing an almost ideal rectangularly shaped spectrum. Nice spectral properties of FM-UWB make it easier to spread the available spectrum into channels. The downside of continuous transmission is that the receiver and transmitter cannot be duty cycled at the symbol level (at least not without a penalty in sensitivity). When it comes to the radio implementation, the FM-UWB generally requires a simpler architecture that consequently allows for lower peak power consumption. The FM-UWB will be studied in detail in the following chapters, while the remainder of this section presents main characteristics of the IR-UWB.

2.3.1 Impulse Radio UWB

As explained, the IR-UWB uses a sequence of pulses to convey information. However, the pulses themselves can be modulated in different ways. Just like in the case of NB radios OOK can be used. The presence of a pulse then corresponds to one symbol and the absence to the other. A second way to modulate pulses is to use PPM, where two or more time windows are associated to different symbols. The receiver then detects energy in each window in order to find the pulse and decode the symbol. Both of these approaches are widely used, mainly owing to the simplicity of the receiver.

The LO generation can be completely avoided and a simple envelope detector used for demodulation. The advantage of the PPM is that a pulse is present for every symbol, the energy is then integrated in the two timing windows (for binary PPM) and simply compared, while some sort of threshold search and track algorithm must be used for OOK. The advantage of the OOK is lower PSD for the same energy per pulse and pulse repetition ratio (PRR). A third option is to use PSK for pulse modulation. However, this modulation was rarely used in combination with the IR-UWB as it requires a coherent demodulator which makes the receiver more complex. The last option found in literature is the FSK. Some of the spectral properties in this case can be controlled through the FSK modulation index. When it comes to performance it is equivalent to PPM, except that instead in time the integration is done in two frequency bands. Although complexity is generally similar to that of OOK and PPM [65], this type of modulation was rarely used in the literature.

Unlike with most NB radios, the output power of UWB radios is quite constrained. This is done to prevent the UWB radios from interfering with higher priority spectrum users (WiFi, BT etc.). The average output PSD of a UWB device must not exceed -41.3 dBm/MHz. This imposes a limit on the energy per pulse and the pulse repetition rate. In order to increase the link budget the output power cannot be simply increased, the PRR must be scaled accordingly in order to avoid violating the spectrum mask, effectively imposing the limit on the maximum data rate. Care should also be taken when interpreting sensitivity of IR-UWB receivers. Sensitivity is defined as the minimum power of the signal at the receiver input for which a certain bit error rate (BER) can be achieved. For low power receivers the required BER is usually 10^{-3}. For NB radios the average signal power doesn't change with the symbol duration (or data rate) because the signal is continuous. However, the average power of the IR-UWB signal will reduce if the symbol duration decreases. Even though the sensitivity improves, the total range and the link budget stay the same. Since the average transmit power decreases as well, it is possible to increase energy per pulse on the transmitter side without violating the spectrum mask. This is sometimes referred to as the duty cycle gain. For this reason some authors report sensitivity normalized to data rate. Another thing to note is that for the case of IR-UWB increasing symbol duration doesn't increase energy per bit, so long as the pulse duration is unchanged.

A typical IR-UWB transmitter is shown in Figure 2.6. Even though the whole PLL is shown it is not necessary to use it. Given the large bandwidth of the signal, carrier frequency doesn't need to be precise. Therefore, a free running oscillator, such as a calibrated ring oscillator, is sufficient. In some

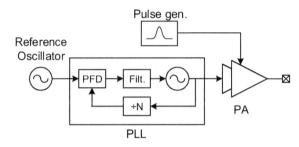

Figure 2.6 IR-UWB transmitter block diagram.

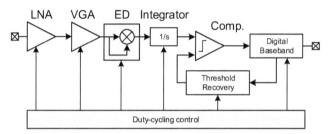

Figure 2.7 IR-UWB OOK/PPM receiver block diagram.

cases even a tuned delay circuit can be used to generate a pulse. Precise frequency control is mainly needed if FSK is used for pulse modulation. Amplitude control in the PA is not necessary, but is often used to improve the spectrum shape. Gaussian pulse shaping is typically used and since the Fourier transform of a Gaussian pulse is also Gaussian, the resulting spectrum will have a Gaussian shape. Proper pulse shaping helps not only to lower the side lobes, but also to make the spectrum more flat, allowing higher output power without violating the mask. Amplitude modulation adds a degree of complexity and makes the transmitter less energy efficient, but the benefits often outweigh the downsides. The fact that the pulse is usually much shorter than the symbol duration can be exploited to duty cycle the transmitter. The circuits can easily be kept off almost all the time and only switched on to transmit a pulse, drastically reducing the average power consumption of the transmitter. The requirement is simply to optimize circuits for fast start-up.

A generic receiver architecture of an OOK/PPM modulated IR-UWB receiver is given in Figure 2.7. The envelope detector based architecture resembles that of the standard WU receiver. The main difference is that all the circuits need to support a much larger bandwidth. As a general rule, these circuits consume a larger amount of power in order to provide the same gain and noise figure as the narrow-band circuits. This is the main disadvantage of UWB radios compared to NB radios.

Using the same reasoning as for the transmitter, one might also think about duty cycling the receiver. The problem on the receiver side is that it is not a priori known when the pulses are going to arrive. Before duty cycling can be applied, the receiver must synchronize to the incoming pulses. Usually the preamble of the packet is used to achieve synchronization so that the receiver can be duty cycled while receiving the data payload. Such a scheme is useful only if the data payload is sufficiently long to provide significant overall power savings. A different scheme is proposed in [43], where the receiver is always duty cycled. To accommodate this the preamble is done in a specific way – the 31 bit synchronization sequence is repeated 19 times. The receiver is integrating the envelope of the signal over a certain time window. Since the integration time is much shorter than the symbol time, it might happen that there is no overlap between the transmitted pulses and the receiver integration time. If this occurs the time reference of the receiver is shifted. The process continues until the receiver correctly detects the synchronization sequence. After that the correct pulse timing is known for the data payload. The downside of this approach is that the synchronization sequence is longer and that in the worst case the receiver will spend at least the same amount of energy for synchronization as in the case without duty cycling. The average energy needed for synchronization should still be lower with the proposed data format from [43].

Selected IR-UWB receivers are compared in Table 2.3. A large variety of performances can be observed, with the data rates going from 30 kb/s up to 500 Mb/s. The implementations interesting for WSN applications are generally in the region between 100 kb/s and 1 Mb/s. Although high speed IR-UWB radios come with highest efficiencies (even compared to NB WU radios), high peak power consumption might pose a problem in a battery powered device, as large peak current might damage the battery. Moderate data rates with low to moderate consumption are a preferred choice in such scenarios. Most implementations are using OOK and PPM, and some implementations come with a modified custom modulation in order to improve performance. The S-OOK modulation, introduced in [41], stands for synchronous OOK. In this type of modulation a leading pulse is introduced for each symbol. This simplifies synchronization on the receiver side as the position of the information carrying pulse is known with respect to the leading pulse. It also allows to remove a precise timing reference from the receiver, since it simply needs to lock to reference pulses (frequency drift

Table 2.3   Performance comparison of IR-UWB receivers

Ref.	Sens. [dBm]	Cons. [mW]	Eff. [pJ/b]	Data Rate [Mb/s]	Freq. [GHz]	BW [MHz]	Mod.
[38]	−59	13.3	26.6	500	7.875	1250	OOK
[39]	−50	13	130	100	4	2000	OOK
[40]	−76.5	0.75	375	2	4.35	500	OOK
[41]	−60	1.64	1640	1	3.8	500	S–OOK
[42]	−74	21	21000	1	3–4	<1000	FH–OOK
[43]	−70	4.2	840	5	7.25–8	1000	PPM
[43]	−87	4.2	4200	0.1	7.25–8	1000	PPM
[44]	−65.8	30.5	305	100	3.1–4.9	1500	RA–OOK
[45]	−67	0.406	1230	0.03	9.8	1000	PPM
[46]	−87*	0.110	800	0.14	3.5–4.5	500	OOK
[47]	−42**	18	180	10	3–5	500	OOK
[48]	−76	22.5	1400	16	3–5	500	PPM/OOK
[49]	−70	6.6	320	20.8	5	1000	OOK
[50]	−99*	35.8	2500	<16.7	3–5	500	PPM
[51]	−88	1.3	1300	1	4.85	600	OOK

*At 100 kb/s.
**Estimated.

of an on-chip oscillator should be negligible over one symbol period). The downside of S-OOK is the increased PRR, as more than one pulse is now needed per symbol, effectively decreasing the maximum allowed transmit power. The transceiver described in [42] uses frequency hopping (FH) OOK. each transmitted pulse is shifted in frequency, spreading the spectrum of the transmitted signal and lowering the PSD. In this way a higher output power can be achieved and link budget can be extended while conforming to the PSD limits. The RA-OOK (random alternate OK) presented in [44] is a simple modification of OOK that improves the spectral properties of the signal. By randomizing phase of the pulses the spectral lines in the output spectrum that might appear due to periodicity of the pulses are avoided. In the same implementation symbol level duty cycling is replaced with a burst level duty cycling. The authors have recognized the difficulty of performing duty cycling at a higher PRR, typically above 10 MHz. This is equivalent to sending very small packets of data and performing packet level duty cycling as is generally done. Aggressive duty cycling of the entire chain is used in the receiver from [40]. Even the bias circuits are on during only 3.9% of the time. Unfortunately, there is power overhead associated to the duty cycle control circuits that cannot be avoided, but the receiver still achieves one the best efficiencies among the medium data rate devices. Similarly, symbol level duty cycling is employed in all the receivers from [45–49]. The implementation reported in [50] maintains similar efficiency over three decades of data rates, with leakage currents degrading performance at very low data rates. The transmitter from [49] uses digital delay cells for pulse generation. These delay cells are calibrated using a delay locked loop and a crystal oscillator. Once calibrated, the transceiver can practically operate without a crystal reference. An effort was also made at a higher level to synchronize IR-UWB radios at a protocol level in order to avoid the use of crystal oscillators [46].

All together, the IR-UWB seems like a promising technique for WSN and IoT types of applications offering a wide range of performance while maintaining very low energy per bit. The main question currently is how to maximize savings from symbol level duty cycling and adapt protocols to minimize dependence on crystal clocks in order to allow further miniaturization of future sensor nodes. Concepts have been proven to work, but the technology still needs to mature before it can be adopted by the industry.

2.4 Concluding Remarks

Many different low power radios have been been developed, the aim of this chapter is to present those that target a similar applications space as the FM-UWB, meaning similar data rate and power consumption. All the radios presented in the previous sections are shown together in Figure 2.8 that compares the data rate and power consumption. Three regions can be

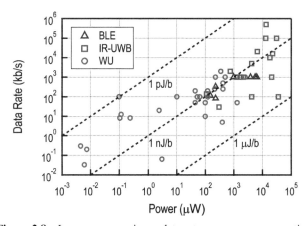

Figure 2.8 Low power receivers, data rate vs. power consumption.

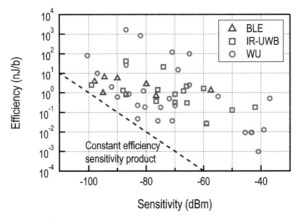

Figure 2.9   Low power receivers, sensitivity vs. efficiency.

differentiated, WU receivers consume the least, and occupy the left part of the graph. The IR-UWB receivers consume the highest peak power and are mainly located in the right part of the graph. Finally, the BLE receivers are located between the two. A general trend that higher data rates require higher power consumption can be easily observed, if we focus on constant efficiency lines, we can see that most implementations are located close the 1 nJ/b, with a few exceptions approaching 1 pJ/b.

Other than data rate, power consumption is also affected by sensitivity. Sensitivity itself is affected by the data rate as decreasing the data rate increases symbol duration, which consequently increases energy per bit and therefore sensitivity. At the same time sensitivity is dependent on the receiver noise figure. Lowering the NF to improve sensitivity requires more power from the LNA, which is often the dominant block in a low power receiver. Efficiency, defined as the ratio of power consumption and data rate, captures both of these effects when compared against sensitivity. The same receivers are shown in efficiency vs. sensitivity graph in Figure 2.9. Here, a product of efficiency and sensitivity can be regarded as a kind of figure of merit (albeit with some reserve). Once the power consumption is normalized to the symbol duration, the three regions vanish and regardless of the type of receiver, similar performance seems to be achievable. To improve sensitivity we must either improve the receiver NF, which costs additional power, or communicate more slowly, and both means increase energy per bit.

This is off course just a part of the whole story. Many other aspects play an important role in receiver performance. Carrier frequency is one of them, as typically circuit consumption grows with frequency. Looking at an LNA as an example, achieving the same gain and NF requires more power at higher frequencies. Better, but also larger components are available at lower frequencies, allowing better matching networks. Most of the wake up receivers operate in the sub-GHz range, which contributes to their consumption. Such low consumption cannot be achieved in the 4–10 GHz range where most of the IR-UWB receivers are operating. The second important aspect is the dynamic range, or linearity of the receiver. As an example, the consumption of an LNA can be decreased by lowering its supply voltage. This results in lower power consumption, but it also reduces headroom and consequently the compression point. In an environment where multiple devices are present, a signal of one radio will interfere with another. If a receiver is easily saturated by the interfering signal it will be impossible to use it in such scenarios.

To understand all advantages of a certain communication scheme one must look beyond just a few parameters. When compared to standard modulations, FM-UWB is inherently penalized in terms of to sensitivity. For the same energy per bit, the BER of FM-UWB is by construction worse than that of the standard FSK. Instead it offers more robustness in realistic environments (in the presence of multipath fading and interferers) and provides a high potential for miniaturization. Both of these properties are highly important in the IoT and WSN applications, and guarantee that the FM-UWB will find its place among the existing radios.

References

[1] J. Gubbi, R. Buyya, S. Marusic, and M. Palaniswami, "Internet of things (iot): A vision, architectural elements, and future directions," *Future Generation Computer Systems*, vol. 29, no. 7, pp. 1645–1660, 2013.

[2] *SX1276/77/78/79 Datasheet*, Semtech, 2019.

[3] T. Sano, M. Mizokami, H. Matsui, K. Ueda, K. Shibata, K. Toyota, T. Saitou, H. Sato, K. Yahagi, and Y. Hayashi, "A 6.3 mW BLE transceiver embedded RX image-rejection filter and TX harmonic-suppression filter reusing on-chip matching network," in *2015 IEEE International Solid-State Circuits Conference – (ISSCC) Digest of Technical Papers*, Feb. 2015, pp. 1–3.

[4] B. Razavi, *RF microelectronics*. Prentice hall New Jersey, 2011.

[5] A. Sai, H. Okuni, T. T. Ta, S. Kondo, T. Tokairin, M. Furuta, and T. Itakura, "A 5.5 mW ADPLL-Based Receiver With a Hybrid Loop Interference Rejection for BLE Application in 65 nm CMOS," *IEEE*

Journal of Solid-State Circuits, vol. 51, no. 12, pp. 3125–3136, Dec. 2016.

[6] Y. Liu, X. Huang, M. Vidojkovic, A. Ba, P. Harpe, G. Dolmans, and H. d. Groot, "A 1.9 nj/b 2.4 ghz multistandard (Bluetooth Low Energy/Zigbee/IEEE802.15.6) transceiver for personal/body-area networks," in *2013 IEEE International Solid-State Circuits Conference Digest of Technical Papers*, Feb. 2013, pp. 446–447.

[7] *PSoC 63 with BLE Datasheet*, Cypress, 2019.

[8] *DA14580 Datasheet*, Dialog Semiconductor, 2014.

[9] *RSL10 Datasheet*, ON Semiconductor, 2018.

[10] A. H. M. Shirazi, H. M. Lavasani, M. Sharifzadeh, Y. Rajavi, S. Mirabbasi, and M. Taghivand, "A 980 μW 5.2 dB-NF current-reused direct-conversion bluetooth-low-energy receiver in 40 nm CMOS," in *2017 IEEE Custom Integrated Circuits Conference (CICC)*, Apr. 2017, pp. 1–4.

[11] M. R. Abdelhamid, A. Paidimarri, and A. P. Chandrakasan, "A −80 dBm BLE-compliant, FSK wake-up receiver with system and within-bit dutycycling for scalable power and latency," in *2018 IEEE Custom Integrated Circuits Conference (CICC)*, Apr. 2018, pp. 1–4.

[12] A. Alghaihab, J. Breiholz, H. Kim, B. Calhoun, and D. D. Wentzloff, "A 150 μW −57.5 dBm-Sensitivity Bluetooth Low-Energy Back-Channel Receiver with LO Frequency Hopping," in *2018 IEEE Radio Frequency Integrated Circuits Symposium (RFIC)*, June 2018, pp. 324–327.

[13] *CC2540 Datasheet*, Texas Instruments, 2013.

[14] *nRF51822 Product Specification*, Nordic Semiconductor, 2014.

[15] *nRF52832 Product Specification*, Nordic Semiconductor, 2017.

[16] *EM 9301 Datasheet*, EM Microelectronic, 2013.

[17] *TC35679 Datasheet*, Toshiba, 2018.

[18] *TC35681 Datasheet*, Toshiba, 2019.

[19] H. Jiang, P. H. P. Wang, L. Gao, P. Sen, Y. H. Kim, G. M. Rebeiz, D. A. Hall, and P. P. Mercier, "A 4.5 nW wake-up radio with −69 dbm sensitivity," in *2017 IEEE International Solid-State Circuits Conference (ISSCC)*, Feb. 2017, pp. 416–417.

[20] C. Hambeck, S. Mahlknecht, and T. Herndl, "A 2.4 μW Wake-up Receiver for wireless sensor nodes with −71 dbm sensitivity," in *2011 IEEE International Symposium on Circuits and Systems (ISCAS)*, May 2011, pp. 534–537.

[21] K.-W. Cheng, X. Liu, and M. Je, "A 2.4/5.8 GHz 10 μW wake-up receiver with −65/−50 dBm sensitivity using direct active rf detection,"

in *Solid State Circuits Conference (A-SSCC), 2012 IEEE Asian*, Nov. 2012, pp. 337–340.

[22] S. Drago, D. Leenaerts, F. Sebastiano, L. Breems, K. Makinwa, and B. Nauta, "A 2.4 GHz 830 pJ/bit duty-cycled wake-up receiver with -82 dBm sensitivity for crystal-less wireless sensor nodes," in *Solid-State Circuits Conference Digest of Technical Papers (ISSCC), 2010 IEEE International*, Feb. 2010, pp. 224–225.

[23] J. Moody, P. Bassirian, A. Roy, N. Liu, N. S. Barker, B. H. Calhoun, and S. M. Bowers, "Interference Robust Detector-First Near-Zero Power Wake-Up Receiver," *IEEE Journal of Solid-State Circuits*, pp. 1–14, 2019.

[24] P. P. Wang, H. Jiang, L. Gao, P. Sen, Y. Kim, G. M. Rebeiz, P. P. Mercier, and D. A. Hall, "A 6.1-nW Wake-Up Receiver Achieving -80.5-dBm Sensitivity Via a Passive Pseudo-Balun Envelope Detector," *IEEE Solid-State Circuits Letters*, vol. 1, no. 5, pp. 134–137, May 2018.

[25] H. Milosiu, F. Oehler, M. Eppel, D. Fruehsorger, S. Lensing, G. Popken, and T. Thoenes, "A 3- μW 868-MHz wake-up receiver with -83 dBm sensitivity and scalable data rate," in *2013 Proceedings of the ESSCIRC (ESSCIRC)*, Sep. 2013, pp. 387–390.

[26] N. M. Pletcher, S. Gambini, and J. Rabaey, "A 52 μW wake-up receiver with -72 dBm sensitivity using an uncertain-IF architecture," *IEEE Journal of Solid-State Circuits*, vol. 44, no. 1, pp. 269–280, Jan. 2009.

[27] N. Roberts and D. Wentzloff, "A 98 nw wake-up radio for wireless body area networks," in *2012 IEEE Radio Frequency Integrated Circuits Symposium (RFIC)*, June 2012, pp. 373–376.

[28] S. Oh, N. E. Roberts, and D. D. Wentzloff, "A 116 nw multi-band wake-up receiver with 31-bit correlator and interference rejection," in *Proceedings of the IEEE 2013 Custom Integrated Circuits Conference*, Sep. 2013, pp. 1–4.

[29] J. Pandey, J. Shi, and B. Otis, "A 120 μw mics/ism-band fsk receiver with a 44 μw low-power mode based on injection-locking and 9x frequency multiplication," in *2011 IEEE International Solid-State Circuits Conference*, Feb. 2011, pp. 460–462.

[30] N. E. Roberts, K. Craig, A. Shrivastava, S. N. Wooters, Y. Shakhsheer, B. H. Calhoun, and D. D. Wentzloff, "26.8 A 236 nW -56.5 dBm-sensitivity bluetooth low-energy wakeup receiver with energy harvesting in 65 nm CMOS," in *2016 IEEE International Solid-State Circuits Conference (ISSCC)*, Jan. 2016, pp. 450–451.

[31] B. Otis, Y. H. Chee, and J. Rabaey, "A 400 μW-RX, 1.6 mW-TX super-regenerative transceiver for wireless sensor networks," in *Solid-State Circuits Conference, 2005. Digest of Technical Papers. ISSCC. 2005 IEEE International*, Feb. 2005, pp. 396–606 Vol. 1.

[32] X. Huang, P. Harpe, G. Dolmans, and H. de Groot, "A 915 MHz ultra-low power wake-up receiver with scalable performance and power consumption," in *ESSCIRC (ESSCIRC), 2011 Proceedings of the*, Sep. 2011, pp. 543–546.

[33] D. Daly and A. Chandrakasan, "An Energy-Efficient OOK Transceiver for Wireless Sensor Networks," *IEEE Journal of Solid-State Circuits*, vol. 42, no. 5, pp. 1003–1011, May 2007.

[34] T. Abe, T. Morie, K. Satou, D. Nomasaki, S. Nakamura, Y. Horiuchi, and K. Imamura, "An ultra-low-power 2-step wake-up receiver for IEEE 802.15.4g wireless sensor networks," in *2014 Symposium on VLSI Circuits Digest of Technical Papers*, June 2014, pp. 1–2.

[35] J. Ayers, K. Mayaram, and T. Fiez, "An Ultralow-Power Receiver for Wireless Sensor Networks," *IEEE Journal of Solid-State Circuits*, vol. 45, no. 9, pp. 1759–1769, Sep. 2010.

[36] C. Salazar, A. Cathelin, A. Kaiser, and J. Rabaey, "A 2.4 ghz interferer-resilient wake-up receiver using a dual-if multi-stage n-path architecture," *IEEE Journal of Solid-State Circuits*, vol. 51, no. 9, pp. 2091–2105, Sep. 2016.

[37] L. Jae-Seung, K. Joo-Myoung, L. Jae-Sup, H. Seok-Kyun, and L. Sang-Gug, "13.1 A 227 pJ/b −83 dbm 2.4 GHz multi-channel OOK receiver adopting receiver-based FLL," in *Solid- State Circuits Conference – (ISSCC), 2015 IEEE International*, Feb. 2015, pp. 1–3.

[38] S. Geng, D. Liu, Y. Li, H. Zhuo, W. Rhee, and Z. Wang, "A 13.3 mW 500 Mb/s IR-UWB Transceiver With Link Margin Enhancement Technique for Meter-Range Communications," *IEEE Journal of Solid-State Circuits*, vol. 50, no. 3, pp. 669–678, Mar. 2015.

[39] L. Xia, K. Shao, H. Chen, Y. Huang, Z. Hong, and P. Y. Chiang, "0.15-nJ/b 3–5-GHz IR-UWB System With Spectrum Tunable Transmitter and Merged-Correlator Noncoherent Receiver," *IEEE Transactions on Microwave Theory and Techniques*, vol. 59, no. 4, pp. 1147–1156, Apr. 2011.

[40] B. Vigraham and P. Kinget, "A self-duty-cycled and synchronized UWB receiver SoC consuming 375 pJ/b for −76.5 dBm sensitivity at 2 Mb/s," in *2013 IEEE International Solid-State Circuits Conference Digest of Technical Papers*, Feb. 2013, pp. 444–445.

[41] M. Crepaldi, C. Li, J. R. Fernandes, and P. R. Kinget, "An Ultra-Wideband Impulse-Radio Transceiver Chipset Using Synchronized-OOK Modulation," *IEEE Journal of Solid-State Circuits*, vol. 46, no. 10, pp. 2284–2299, Oct. 2011.

[42] D. Liu, X. Ni, R. Zhou, W. Rhee, and Z. Wang, "A 0.42-mW 1-Mb/s 3- to 4-GHz Transceiver in 0.18- $\mu \textm$ CMOS With Flexible Efficiency, Bandwidth, and Distance Control for IoT Applications," *IEEE Journal of Solid-State Circuits*, vol. 52, no. 6, pp. 1479–1494, June 2017.

[43] S. Solda, M. Caruso, A. Bevilacqua, A. Gerosa, D. Vogrig, and A. Neviani, "A 5 Mb/s UWB-IR Transceiver Front-End for Wireless Sensor Networks in 0.13$\mu\hboxm$CMOS," *IEEE Journal of Solid-State Circuits*, vol. 46, no. 7, pp. 1636–1647, July 2011.

[44] R. Vauche, E. Muhr, O. Fourquin, S. Bourdel, J. Gaubert, N. Dehaese, S. Meillere, H. Barthelemy, and L. Ouvry, "A 100 MHz PRF IR-UWB CMOS Transceiver With Pulse Shaping Capabilities and Peak Voltage Detector," *IEEE Transactions on Circuits and Systems I: Regular Papers*, vol. 64, no. 6, pp. 1612–1625, June 2017.

[45] J. K. Brown, K. Huang, E. Ansari, R. R. Rogel, Y. Lee, and D. D. Wentzloff, "An ultra-low-power 9.8 GHz crystal-less UWB transceiver with digital baseband integrated in 0.18 μm BiCMOS," in *2013 IEEE International Solid-State Circuits Conference Digest of Technical Papers*, Feb. 2013, pp. 442–443.

[46] X. Y. Wang, R. K. Dokania, and A. B. Apsel, "A Crystal-Less Self-Synchronized Bit-Level Duty-Cycled IR-UWB Transceiver System," *IEEE Transactions on Circuits and Systems I: Regular Papers*, vol. 60, no. 9, pp. 2488–2501, Sep. 2013.

[47] A. Ebrazeh and P. Mohseni, "A 14 pJ/pulse-TX, 0.18 nJ/b-RX, 100 Mbps, channelized, IR-UWB transceiver for centimeter-to-meter range biotelemetry," in *Proceedings of the IEEE 2014 Custom Integrated Circuits Conference*, Sep. 2014, pp. 1–4.

[48] D. C. Daly, P. P. Mercier, M. Bhardwaj, A. L. Stone, Z. N. Aldworth, T. L. Daniel, J. Voldman, J. G. Hildebrand, and A. P. Chandrakasan, "A Pulsed UWB Receiver SoC for Insect Motion Control," *IEEE Journal of Solid-State Circuits*, vol. 45, no. 1, pp. 153–166, Jan. 2010.

[49] A. Mehra, M. Sturm, D. Hedin, and R. Harjani, "A 0.32 nJ/bit noncoherent UWB impulse radio transceiver with baseband synchronization and a fully digital transmitter," in *2013 IEEE Radio Frequency Integrated Circuits Symposium (RFIC)*, June 2013, pp. 17–20.

[50] F. S. Lee and A. P. Chandrakasan, "A 2.5 nJ/bit 0.65 V Pulsed UWB Receiver in 90 nm CMOS," *IEEE Journal of Solid-State Circuits*, vol. 42, no. 12, pp. 2851–2859, Dec. 2007.

[51] B. Vigraham and P. R. Kinget, "An ultra low power, compact UWB receiver with automatic threshold recovery in 65 nm CMOS," in *2012 IEEE Radio Frequency Integrated Circuits Symposium*, June 2012, pp. 251–254.

[52] J. Long, W. Wu, Y. Dong, Y. Zhao, M. A. T. Sanduleanu, J. F. M. Gerrits, and G. van Veenendaal, "Energy-efficient wireless front-end concepts for ultra lower power radio," in *IEEE Custom Integrated Circuits Conference, 2008. CICC 2008*, Sep. 2008, pp. 587–590.

[53] J. Pandey and B. P. Otis, "A sub-100 μW MICS/ISM band transmitter based on injection-locking and frequency multiplication," *IEEE Journal of Solid-State Circuits*, vol. 46, no. 5, pp. 1049–1058, May 2011.

[54] P. J. A. Harpe, C. Zhou, Y. Bi, N. P. van der Meijs, X. Wang, K. Philips, G. Dolmans, and H. de Groot, "A 26 μw 8 bit 10 ms/s asynchronous sar adc for low energy radios," *IEEE Journal of Solid-State Circuits*, vol. 46, no. 7, pp. 1585–1595, July 2011.

[55] J. Farserotu, J. Baborowski, J. D. Decotignie, P. Dallemagne, C. Enz, F. Sebelius, B. Rosen, C. Antfolk, G. Lundborg, A. Björkman, T. Knieling, and P. Gulde, "Smart skin for tactile prosthetics," in *2012 6th International Symposium on Medical Information and Communication Technology (ISMICT)*, Mar. 2012, pp. 1–8.

[56] *IEEE Standard for Local and Metropolitan Area Networks – Part 15.6: Body Area Networks*, 2012.

[57] M. Verhelst, N. V. Helleputte, G. Gielen, and W. Dehaene, "A reconfigurable, 0.13 μm cmos 110 pJ/pulse, fully integrated IR-UWB receiver for communication and sub-cm ranging," in *2009 IEEE International Solid-State Circuits Conference – Digest of Technical Papers*, Feb. 2009, pp. 250–251,251a.

[58] A. Medi and W. Namgoong, "A high data-rate energy-efficient interference-tolerant fully integrated cmos frequency channelized uwb transceiver for impulse radio," *IEEE Journal of Solid-State Circuits*, vol. 43, no. 4, pp. 974–980, Apr. 2008.

[59] Y. Zheng, Y. Tong, C. W. Ang, Y.-P. Xu, W. G. Yeoh, F. Lin, and R. Singh, "A CMOS carrier-less UWB transceiver for WPAN applications," in *2006 IEEE International Solid State Circuits Conference – Digest of Technical Papers*, Feb. 2006, pp. 378–387.

[60] M. Crepaldi, C. Li, K. Dronson, J. Fernandes, and P. Kinget, "An ultra-low-power interference-robust IR-UWB transceiver chipset using self-synchronizing OOK modulation," in *2010 IEEE International Solid-State Circuits Conference – (ISSCC)*, Feb. 2010, pp. 226–227.

[61] F. Zhang, A. Jha, R. Gharpurey, and P. Kinget, "An agile, ultra-wideband pulse radio transceiver with discrete-time wideband-IF," *IEEE Journal of Solid-State Circuits*, vol. 44, no. 5, pp. 1336–1351, May 2009.

[62] L. Liu, T. Sakurai, and M. Takamiya, "A 1.28 mW 100 Mb/s impulse UWB receiver with charge-domain correlator and embedded sliding scheme for data synchronization," in *2009 Symposium on VLSI Circuits*, June 2009, pp. 146–147.

[63] R. Liu, B. R. Carlton, S. Pellerano, F. Sheikh, D. S. Vemparala, A. Ali, and V. S. Somayazulu, "A 264-μW 802.15.4a-Compliant IR-UWB Transmitter in 22 nm FinFET for Wireless Sensor Network Application," in *2018 IEEE Radio Frequency Integrated Circuits Symposium (RFIC)*, June 2018, pp. 164–167.

[64] X. Wang, Y. Yu, B. Busze, H. Pflug, A. Young, X. Huang, C. Zhou, M. Konijnenburg, K. Philips, and H. D. Groot, "A meter-range UWB transceiver chipset for around-the-head audio streaming," in *2012 IEEE International Solid-State Circuits Conference*, Feb. 2012, pp. 450–452.

[65] D. Barras, "A low power impulse radio ultra-wideband cmos radio-frequency transceiver," PhD dissertation, ETH Zurich, 2010.

3

FM-UWB as a Low-Power, Robust Modulation Scheme

3.1 Introduction

Ultra-wideband (UWB) systems were originally intended to provide robust, low-cost, low-complexity and low power wireless solutions for localization and communication. The first UWB systems were based on a time domain approach, they used a very short pulse to carry the information. Initially, they were used in radar systems, where pulse duration translated into spatial resolution. When used for communications, these pulses could be modulated using one of the standard approaches, such as OOK, PPM, PSK or FSK. This was impulse radio (IR) UWB, and although it was able to provide robust, high-speed communication, it came at the price of circuit complexity and relatively high peak power consumption. The frequency-modulated (FM) UWB was developed as an easy to implement, complementary solution, preserving robustness and offering low to medium data rates. This analog spread spectrum technique is intended for short to medium range applications that require a reliable communication link, low cost and high degree of integration and miniaturization, and therefore perfectly fits the IoT requirements.

This chapter begins by introducing the fundamentals of FM-UWB, explaining the modulation and demodulation principles and basic transmitter and receiver architectures. Then, the Gerrits' BER approximation is presented and extended to cases with multiple FM-UWB users and narrowband interferers. Finally, possible extensions of standard FM-UWB modulation are briefly discussed, highlighting its potential evolution. In the second part of this chapter, state of the art FM-UWB receivers and transmitters are discussed and analyzed, and a brief summary of their key characteristics is provided. They are also compared to narrowband and IR-UWB radios to point out the advantages and disadvantages of the FM-UWB modulation scheme.

3.2 Principles of FM-UWB

3.2.1 FM-UWB Modulation

The FM-UWB can be seen as an analog spread-spectrum technique. In its basic form it is a double FM modulation. A low modulation index FSK, called a sub-carrier, is followed by a high modulation index FM ($\beta \gg 1$) to achieve large bandwidth. The principle of FM-UWB modulation is shown in Figure 3.1. The resulting FM-UWB signal can be represented as [1]:

$$s_{UWB}(t) = A \cos \left(\omega_c t + \Delta\omega \int_{-\infty}^{t} m(t)\mathrm{d}t \right) = A \cos \left(\omega_c t + \phi(t) \right), \quad (3.1)$$

where ω_c is the center frequency, $\Delta\omega = 2\pi\Delta f$ is the frequency deviation and $m(t)$ is the normalized, FSK modulated sub-carrier. According to definition, to be considered UWB the signal must either exceed 500 MHz or 20% of its center frequency. The bandwidth of the FM signal can be approximated using the Carson's rule [1]:

$$B_{FM} = 2f_m(\beta + 1) = 2(\Delta f + f_m). \quad (3.2)$$

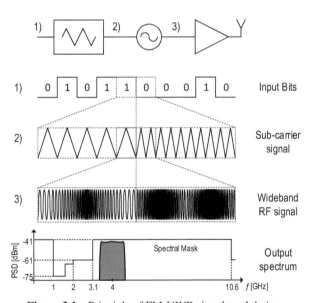

Figure 3.1   Principle of FM-UWB signal modulation.

In the above equation f_m is the maximum frequency in the FSK signal spectrum which depends on the sub-carrier center frequency f_{SC} and the data rate R, according to $f_m = f_{SC} + R$. Spectral properties of the FM-UWB signal depend on the sub-carrier waveform. For an FM signal with modulation index much larger than unity, quasi-stationary approximation is valid and the FM-UWB signal power spectral density (PSD) will be a function of the probability density function (PDF) p_m of $m(t)$ [2]:

$$S_{FM-UWB}(\omega) = \frac{\pi A^2}{2} \left[p_m \left(\frac{\omega - \omega_c}{\Delta \omega} \right) + p_m \left(\frac{\omega + \omega_c}{\Delta \omega} \right) \right]. \qquad (3.3)$$

As long as the sub-carrier frequency is reasonably low (keeping the second FM modulation index high, $\beta \gg 1$), the FM-UWB spectrum will be largely determined by the sub-carrier waveform. For an ideal triangular sub-carrier the FM-UWB spectrum will be flat with a relatively steep roll-off. A steeper roll-off can be achieved by using a sinusoidal sub-carrier, but this results in curved spectrum shape, with peaking at the edges of the band [3]. As a result the maximum transmit power must be lowered in order to comply with the spectral mask. At higher sub-carrier frequencies, or equivalently lower modulation index (practically $\beta < 20$) Equation (3.3) is no longer valid, and good spectral properties of the FM-UWB signal are lost.

Performance of the FM-UWB modulation can be studied using a wideband FM demodulator presented in Figure 3.2. After multiplying the signal s_{UWB} with its delayed version and disregarding the high-frequency components, signal at the output of the demodulator will be given by [1]:

$$s_{dem}(t) = \frac{A^2}{2} \cos(\omega_c \tau + \phi(t) - \phi(t - \tau)). \qquad (3.4)$$

By choosing the time delay equal to an odd multiple N of the quarter period of the carrier center frequency $\tau = NT/4 = N\pi/2\omega_c$

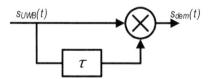

Figure 3.2   Wideband FM demodulator.

$(N = 1, 3, 5, \ldots)$ Equation (3.4) can be written in the following form:

$$s_{dem}(t) = (-1)^{(N+1)/2} \frac{A^2}{2} \sin(\phi(t) - \phi(t - \tau)) \tag{3.5}$$

$$= (-1)^{(N+1)/2} \frac{A^2}{2} \sin\left(\tau \frac{d\phi(t)}{dt}\right) \tag{3.6}$$

$$= (-1)^{(N+1)/2} \frac{A^2}{2} \sin\left(N \frac{\pi \Delta \omega}{2\omega_c} m(t)\right), \tag{3.7}$$

under the assumption that the delay τ is much smaller than the period of the modulating frequency f_m. The bandwidth of the demodulator, herein defined as the frequency range over which the demodulator characteristic is monotonic, depends on N and is given by

$$B_{dem} = f_c \frac{2}{N}. \tag{3.8}$$

A small delay deviation results in offset between the demodulator center frequency and the FM-UWB signal center frequency. This offset will lead to a distortion of the output signal that is dependent on the bandwidth of the signal and the demodulator. It should be noted that the demodulated signal is proportional to the square of the input amplitude (as seen from Equation 3.7). This results in expanded dynamic range of the demodulated signal, for example a $10\,\text{dB}$ variation in the input amplitude causes $20\,\text{dB}$ variation in the demodulated signal amplitude. Furthermore, the signal to noise ratio (SNR) at the demodulator output will be a non-linear function of the input SNR. Based on simplified analysis provided in [1] the SNR at the demodulator output is given by

$$\text{SNR}_{\text{out}} = \frac{B_{RF}}{B_{SC}} \frac{\text{SNR}_{\text{in}}^2}{1 + 4\text{SNR}_{\text{in}}}, \tag{3.9}$$

where SNR_{in} and SNR_{out} represent the signal to noise ratio at the input and the output of the demodulator, respectively. The ratio B_{RF}/B_{SC} is the ratio of the FM-UWB signal bandwidth and sub-carrier bandwidth, and can be seen as a kind of analog processing gain. At the demodulator output the ratio of the energy per bit and the noise PSD is given by

$$(E_b/N_0)_{\text{dem}} = \text{SNR}_{\text{out}} \frac{B_{SC}}{R}, \tag{3.10}$$

where R is the data rate. As shown in [1], assuming a coherent, optimal demodulator and an orthogonal FSK sub-carrier modulation, the BER can be calculated as:

$$P_b = \frac{1}{2}\text{erfc}\left(\sqrt{\frac{(E_b/N_0)_{\text{dem}}}{2}}\right). \tag{3.11}$$

The erfc function is defined as:

$$\text{erfc}(x) = \frac{2}{\sqrt{\pi}}\int_x^\infty e^{\left(-t^2\right)}dt. \tag{3.12}$$

In most cases, however, a practically implemented FSK demodulator will be non-coherent. Even though there is a small performance penalty, a non-coherent demodulator is simpler, easier to implement and consumes less power. For a non-coherent demodulator the probability of error is given by

$$P_b = \frac{1}{2}e^{\frac{(E_b/N_0)_{\text{dem}}}{2}}. \tag{3.13}$$

The penalty for using a non-coherent demodulator is below 1.5 dB.

A comparison between coherent FM-UWB and FSK signals having equal power is given in Figure 3.3. The ratio of energy per bit and noise power spectral density at the input E_b/N_0 is used instead of SNR_{in} in order to provide a fair comparison. This ratio is defined as

$$E_b/N_0 = \text{SNR}_{\text{in}}\frac{B_{RF}}{R}. \tag{3.14}$$

In the given example $B_{RF} = 500\,\text{MHz}$, the sub-carrier modulation index is $\beta_{sub} = 0.5$ (the same modulation index is used for FSK) and R is the

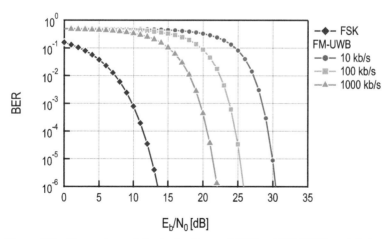

Figure 3.3   Comparison of standard orthogonal FSK and FM-UWB modulation.

data rate. In terms of BER, the FM-UWB is clearly suboptimal compared to standard FSK modulation. This is not surprising considering that the larger signal bandwidth results in higher noise power, that ultimately lowers the input SNR. Part of the lost SNR will be recovered in the process of demodulation owing to the processing gain, however the BER degradation compared to the FSK remains notable. The gap between the two modulations decreases with increasing the FM-UWB data rate, and hence higher data rates should yield better performance. However, higher data rates will also require higher sub-carrier frequencies, which results in loss of the spectral properties of the FM-UWB signal.

Compared to narrowband modulations, FM-UWB is clearly suboptimal in terms of sensitivity. There are however some benefits that are perhaps not apparent at a first glance. The FM-UWB offers robustness against frequency selective fading and interferers. The behavior of FM-UWB signal in a multipath environment has been studied in [4]. It has been shown that even in severe environments performance degradation of FM-UWB is only minor. Owing to the fact that the signal is spread over a very large band, frequency selectivity is not as harmful as it is for narrowband signals (this can be seen as a kind of frequency diversity). The second benefit of using FM-UWB is its inherent robustness to narrowband interferers. Unlike narrowband systems that rely purely on filtering, the FM-UWB provides some inherent interferer rejection. This further implies that it does not require an increase of the receiver complexity or external filters to provide good performance, hence providing higher potential for miniaturization.

3.2.2 Multi-User Communication and Narrowband Interference

In a wireless sensor network, multiple nodes may need to communicate at the same time. One way to resolve this is the time-division multiple access (TDMA), that allocates time slots in which certain nodes can transmit or receive. This approach requires precise synchronization between the nodes, and as the number of nodes in the network grows, the latency increases quickly. Use of other techniques, such as frequency-division multiple-access (FDMA), where different frequencies are allocated to different users, may reduce the overall latency and synchronization requirements. This section studies the behavior of an FM-UWB system in the presence of multiple input signals and is mainly based on the approach presented in [1].

Suppose there are two signals present at the input of the wideband FM demodulator (Figure 3.2) $s_1(t)$ and $s_2(t)$. At the demodulator output the signal will be given by:

$$s_{dem} = s_1(t)s_1(t - \tau) + s_2(t)s_2(t - \tau)$$
$$+ s_1(t)s_2(t - \tau) + s_1(t - \tau)s_2(t). \quad (3.15)$$

Let us assume that the $s_1(t)$ is the FM-UWB signal and the $s_2(t)$ is a narrowband interferer. The component $s_1(t)s_1(t - \tau)$ corresponds to the demodulated sub-carrier. The component $s_2(t)s_2(t - \tau)$ is the FM demodulated narrowband signal. Since its bandwidth is rather small compared to the FM-UWB bandwidth, this component will be located close to dc and can easily be filtered out. It will therefore not influence the sensitivity of the receiver (at least in the ideal case). The last two terms in Equation (3.15) constitute the residual signal that will pollute the useful signal [1]:

$$W(t) = s_1(t)s_2(t - \tau) + s_1(t - \tau)s_2(t). \quad (3.16)$$

The low-frequency terms of the residual signal $W(t)$ will fall within the sub-carrier band, effectively increasing the noise floor of the receiver and lowering sensitivity. Assuming the narrowband signal is located close to the FM-UWB signal center frequency, residual signal will be located at baseband frequencies from 0 to $B_{RF}/2$. If flat spectrum of the residual signal is further assumed, then the signal to interference ratio can be estimated as [1]:

$$\text{SIR} = 20 \log \left(\frac{A_1}{2A_2} \right) - 10 \log \left(\frac{B_{RF}}{2B_{SC}} \right), \quad (3.17)$$

where A_1 and A_2 are the amplitudes of the two input signals. Factor $B_{RF}/2B_{SC}$ is a result of sub-carrier filtering. Interestingly, the amount of interference rejection is proportional to the FM-UWB processing gain.

Multiple FM-UWB signals can be distinguished by assigning different sub-carrier frequencies to different users. This technique will be referred to as the sub-carrier FDMA (SC-FDMA). Assuming that the signals $s_1(t)$ and $s_2(t)$, from Equation (3.15), are the two FM-UWB signals it is clear that the simultaneous demodulation of different FM-UWB signals is possible. The component $s_2(t)s_2(t - \tau)$ will in this case correspond to the second demodulated FM-UWB signal. Provided that the sub-carrier frequency of the second signal is separated from the first, the two can be distinguished and demodulated separately. The principle of multi-user communications using the SC-FDMA is illustrated in Figure 3.4. Multiple signals transmitted from different nodes can be demodulated either by a single node (for example

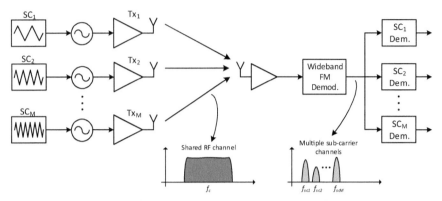

Figure 3.4   FM-UWB multi-user communication.

gathering data from multiple sensors simultaneously), or by different nodes (e.g. to allow isolation of different parts of the network).

Just like in the case of the narrowband interferer, the residual signal will cause sensitivity degradation. Assuming two FM-UWB signals at the input, with aligned center frequencies, Equation (3.16) can be written as:

$$W(t) = \frac{A_1 A_2}{2} \cos\left(\omega_c \tau + \phi_1(t) - \phi_2(t - \tau)\right)$$
$$+ \frac{A_1 A_2}{2} \cos\left(\omega_c \tau + \phi_2(t) - \phi_1(t - \tau)\right) \tag{3.18}$$
$$\approx (-1)^{(N+1)/2} A_1 A_2 \sin\left(\frac{\tau}{2}\left(\frac{d\phi_1(t)}{dt} - \frac{d\phi_2(t)}{dt}\right)\right)$$
$$\times \cos\left(\phi_1(t) + \phi_2(t)\right) \tag{3.19}$$
$$= (-1)^{(N+1)/2} A_1 A_2 \sin\left(\frac{\Delta\omega\tau}{2}\left(m_1(t) - m_2(t)\right)\right)$$
$$\times \cos\left(\phi_1(t) + \phi_2(t)\right). \tag{3.20}$$

The residual signal is proportional to the difference of the two modulating signals multiplied by the factor $\cos\left(\phi_1(t) + \phi_2(t)\right)$, which is a signal that occupies a bandwidth of $B_{RF} = 2\Delta f$. Again, assuming the spectrum of the residual signal is flat, signal-to-interference ratio can be estimated as [1]:

$$\text{SIR} = 20\log\left(\frac{A_1}{2A_2}\right) - 10\log\left(\frac{B_{RF}}{B_{SC}}\right). \tag{3.21}$$

The achievable BER is limited by the SIR. Increasing the number of users, or increasing the difference in power levels between the two users, reduce the SIR and could eventually prevent correct demodulation of the useful signal. The maximum required BER will ultimately limit the tolerable SIR, and subsequently, the number of users or the maximum acceptable power difference.

The analysis conducted by Gerrits in [1, 5] can be extended to the case of multiple FM-UWB users in the presence of noise. Assuming that the delay can be considered relatively small and that the noise autocorrelation function is $R_n(\tau) = 1$ for values of the delay $\tau = N\pi/2\omega_c$, the noise analysis can be simplified while maintaining good accuracy. The result reported in [6] can be generalized for the case of M FM-UWB users. Under the above assumptions the demodulator output signal is given by

$$s_{dem} = (s_1 + s_2 + \cdots + s_M + n)^2 \tag{3.22}$$

$$= \sum_{i=1}^{M} s_i^2 + 2 \sum_{i=1}^{M} \sum_{j=i+1}^{M} s_i s_j + \sum_{i=1}^{M} s_i n + n^2. \tag{3.23}$$

Terms of the form s_i^2 correspond to the demodulated sub-channel i. Terms of the form $2s_i s_j$, $i \neq j$, correspond to the interference among different FM-UWB signals, the number of these terms is $M(M-1)/2$. Finally, following the same reasoning as in [1, 5], and assuming that all the noise and interference terms are independent, the output signal to noise and interference ratio (SNIR) is given by

$$\text{SNIR}_{k,out} = \frac{B_{RF}}{B_{SC}} \frac{S_k^2}{N^2 + 4\sum_{i=1}^{M} S_i N + 4\sum_{i=1}^{M} \sum_{j=i+1}^{M} S_i S_j}, \tag{3.24}$$

where S_i corresponds to the input power of signal s_i and N is the input noise power. For a multi-user environment two cases are of particular importance:

1. Two FM-UWB users of different input power levels
2. M FM-UWB users of equal power levels

For the case of two users, Equation (3.24) reduces to

$$\text{SNIR}_{1,out} = \frac{B_{RF}}{B_{SC}} \frac{S_1^2}{N^2 + 4S_1 N + 4S_2 N + 4S_1 S_2} \tag{3.25}$$

$$= \frac{B_{RF}}{B_{SC}} \frac{\text{SNR}_{1,in}^2}{1 + 4\text{SNR}_{1,in}(1 + \text{SIR}_{in}^{-1}) + 4\text{SNR}_{1,in}^2 \text{SIR}_{in}^{-1}}, \tag{3.26}$$

where $\text{SIR}_{in} = S_1/S_2$. Compared to Equation (3.9) two additional terms exist that depend on the input signal to interferer ratio SIR_{in}. Furthermore, for increasing values of $SNR_{1,in}$ the output signal to noise and interference ratio $SNR_{1,out}$, is no longer limited by noise, but solely by the interference and approaches

$$SNIR_{1,out} = \frac{B_{RF}}{B_{SC}} \frac{SIR_{in}}{4}, \quad \text{for } SIR_{in} \gg 1, \tag{3.27}$$

which is the limit from Equation (3.21). As an example, consider that the FM-UWB signal is used with RF bandwidth $B_{RF} = 500\,MHz$, using a 100 kb/s sub-carrier, with orthogonal FSK and a modulation index of 1 ($B_{SC} = 200\,kHz$). The required SNIR for orthogonal FSK to achieve a BER of 10^{-3} is approximately 13 dB. The maximum difference in power levels between the two FM-UWB signals is then 21 dB.

For the case of M users of equal input power, $S_1 = S_2 = \cdots = S_M = S$, the Equation (3.24) reduces to

$$SNIR_{out} = \frac{B_{RF}}{B_{SC}} \frac{S^2}{N^2 + 4MSN + 2M(M-1)S^2} \tag{3.28}$$

$$= \frac{B_{RF}}{B_{SC}} \frac{SNR_{in}^2}{1 + 4MSNR_{in} + 2M(M-1)SNR_{in}^2}. \tag{3.29}$$

Again, if the signal power is sufficiently higher than the noise power, the output $SNIR_{out}$ becomes a function of FM-UWB signal bandwidth and the number of users:

$$SNIR_{out} = \frac{1}{2M(M-1)} \frac{B_{RF}}{B_{SC}}, \quad \text{for } SIR_{in} \gg 1. \tag{3.30}$$

The above equation can be used to determine the maximum achievable number of users, for a given minimum required signal to noise and distortion ratio. For example, assuming the same system parameters as above ($B_{RF} = 500\,MHz$, $R = 100\,kb/s$, $B_{SC} = 200\,kHz$), the maximum number of equal power users is 16. In both described cases FM-UWB signal bandwidth can be increased in order to increase the achievable $SNIR_{out}$.

The limits predicted by Equations (3.27) and (3.30) assume an ideal system since they only take into account a limited number of effects. In practical systems, these limits are upper bounds and will be difficult to achieve. The above analysis only considers an ideal AWGN channel, with a perfectly flat frequency characteristic. In reality, this will never be the case. Part of the channel transfer function will come from the transmitter and receiver, and part will come from the wireless channel (e.g. due to multi-path propagation). Intuitively, one can see the FM-UWB signal as a carrier that slowly

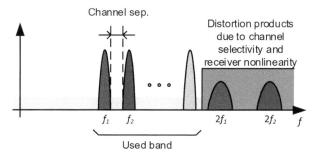

Figure 3.5 FSK sub-channel frequency allocation and limits due to distortion.

moves across a broad frequency range. Since the equivalent channel transfer function is not constant, the amplitude will vary with the instantaneous carrier frequency. Assuming that the channel does not change with time, amplitude will be a periodic function, with the period of the sub-carrier. Even if the wideband FM demodulator is perfect, these amplitude variations will result in the appearance of harmonics. Aside from the channel transfer function, the harmonics will also appear as a product of non-linearities in the receiver chain. Finally, these harmonics will limit the useful sub-carrier band to one octave. If $f_{SC,min}$ is the minimum sub-carrier frequency, then spectrum above $2f_{SC,min}$ will be corrupted by the second and higher order components. The quality of any useful signal at frequencies above $2f_{SC,min}$ would, therefore, be degraded by the harmonics of other sub-carriers, effectively preventing correct demodulation (Figure 3.5). With one octave limit for the sub-carrier band, the lowest sub-carrier frequency $f_{SC,min} = 1$ MHz, and 200 kHz wide FSK channels, the number of sub-carrier channels that can be accommodated is 5. This number can be increased by increasing $f_{SC,min}$. In principle, the effect of channel transfer function can be canceled out by equalization of the input FM-UWB signal. However, equalization techniques are complex, they would drastically increase power consumption of the receiver, and as such are not suited for low power systems.

In the previous example, it was assumed that the FSK channels can be placed adjacent to each other. This is impractical for two reasons. The first reason is that the undesired channels must be filtered out before the final FSK demodulation. Because of the finite quality factor of the filter, some spacing must be introduced between the channels. The second reason is interference among adjacent FSK channels. Theoretically the spectrum of the FSK signal is infinitely wide. Although largest portion of the channel power is located inside the band defined by Carson's rule, part of the spectrum will leak

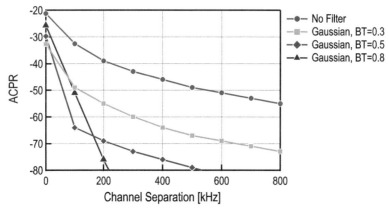

Figure 3.6   ACPR for filtered and non-filtered FSK signal, as a function of channel separation (100 kb/s data rate, modulation index 1).

to side channels and interfere with adjacent users. This effect is quantified by the adjacent channel power ratio (ACPR), and is defined as the ratio of the power inside the channel to the power in the adjacent channel. The ACPR generally depends on the type of modulation, pulse shaping filter and transmitter non-linearity. In the case of FSK modulation, typically Gaussian pulse shaping is used. The shape of the Gaussian pulse is determined by the bandwidth-time (BT) parameter, defined as the ratio between the 3 dB filter bandwidth and data rate. Decreasing the BT parameter results in more compact spectrum, but increases the inter-symbol interference as the pulse duration increases (over several bit periods). ACPR as a function of channel separation, for different values of the BT parameter, is given in Figure 3.6. Although filtering can be used to reduce interference, this was rarely done in reported FM-UWB implementations. The reason is that it adds complexity on both transmitter and receiver sides, and since multi-user communication with FM-UWB has rarely been explored it was not needed. Interference among channels can always be decreased by increasing the channel separation, but this also reduces the number of available FSK channels. For a system with $B_{RF} = 500 \, \text{MHz}$, $R = 100 \, \text{kb/s}$, $B_{SC} = 200 \, \text{kHz}$, the required SNIR of the FSK signal to achieve a BER of 10^{-3} is 13 dB. If a channel separation of 100 kHz is used, with no filtering, then the adjacent channel power can be at most 20 dB above the desired channel power. This will correspond to 10 dB difference in power between the two FM-UWB signals. For this particular case, it is the ACPR that will limit the maximum tolerable power difference between the two users and not the interference from the residual signal (Equation (3.27)).

Additional constraints may come from the receiver non-linearity and limited dynamic range. Due to the quadratic demodulator characteristic, the dynamic range requirements are higher for the circuits following the wide-band FM demodulator. If one of the FSK signals is sufficiently strong it may saturate the circuits causing suppression of weaker FSK signals (FM capture effect). Since there is typically a trade-off between power and dynamic range in amplifiers, a larger acceptable power difference between the received signals will come at the cost of increased power consumption.

Different choice with respect to the system parameters leads to different performance in terms of complexity, sensitivity, data rate, number of chan-nels and power consumption. By modifying the RF bandwidth, sub-carrier frequencies, dynamic range etc., it is possible to perform various trade-offs and to optimize the FM-UWB transceiver according to the specific needs of the system.

3.2.3 Beyond Standard FM-UWB

The FM-UWB modulation was originally intended as double FM modulation, where a low modulation index FSK is followed by a large modulation index FM. It is an optional mode in the IEEE 802.15.6 standard for wireless body area networks [7]. According to the UWB PHY specifications, two modu-lations are supported; IR-UWB as mandatory and FM-UWB as an optional mode. For FM-UWB, the data rate is set to 250 kb/s, using a continuous phase (CP) FSK modulation, centered at 1.5 MHz, with a frequency deviation of 250 kHz. A Gaussian filter is used for pulse shaping with the BT parameter set to 0.8. For the sub-carrier waveform, either a triangular, a sawtooth, or a sine waveforms are allowed.

Strict standard definitions do not allow different sub-carrier frequencies, higher or lower data rates, or multi-user communication. The lack of flexi-bility limits the use of FM-UWB in WBAN applications, and does not allow FM-UWB to reach its full potential. In general, the sub-carrier modulation does not need to be limited to 250 kb/s 2-FSK. Speed and modulation order could be modified according to the channel conditions (a less frequency selective channel allows higher data rates). A transmitter implementing a data rate of 1 Mb/s has been reported in [8], that demonstrates the feasibility of moving to higher data rates. Furthermore, higher order FSK can be explored, such as 4-FSK and 8-FSK, allowing to further boost communication speed.

One such transmitter is reported in [9]. Finally, it would also be possible to use PSK modulations without affecting the good spectral properties of FM-UWB (note that standard PSK modulation requires a coherent SC demodulator). Other variations are possible, and one example is the Chirp-UWB (C-UWB) modulation [10], that is a trade-off between FM-UWB and IR-UWB. Instead of continuous frequency sweep, a single up or down chirp is transmitted depending on the input bit. The duration of the chirp is much lower than the symbol duration and allows duty cycling of the transceiver at a symbol level, thus saving power. At the same time the duration of the pulse is much longer than in the case of IR-UWB and does not require precise synchronization. One downside of C-UWB is that the good spectral properties of the FM-UWB signal are lost.

A minor modification of a standard FM-UWB signal can be used to enable simultaneous transmission on multiple sub-channels. Instead of using a single FSK sub-channel, multiple sub-channels can be summed, and the resulting signal used to modulate the RF carrier. This would allow a single transmitter to transmit different messages to multiple receivers at the same time. The concept is shown in Figure 3.7. In order to preserve the same frequency deviation, if M sub-channels are used, sub-carrier signals are scaled by a factor $1/M$. The example for two sub-carriers is shown in Figure 3.8. Unfortunately, the flat spectrum of the transmitted signal is lost and, as a consequence, transmit power will have to be decreased in order to maintain the signal below the spectral mask defined for the UWB band. The spectrum will take the shape of the PDF of the modulating signal (as shown by Equation (3.3)) which is in this case an average of the two sub-carrier signals, and is no longer a triangular waveform. The exact shape of the resulting sum of sub-carrier signals will

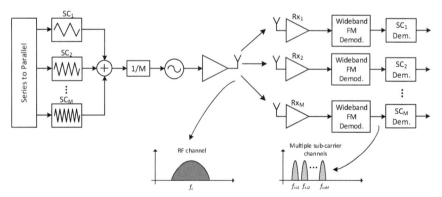

Figure 3.7   FM-UWB multi-channel broadcast.

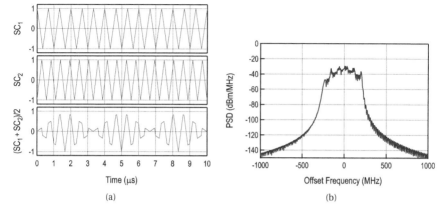

Figure 3.8 Example of transmission on two channels, time domain sub-carrier signal (a) and transmitted signal spectrum (b).

depend on the number of sub-channels, their frequencies and initial phases. The BER calculation can be extended to the case of M sub-channels. The only difference compared to standard FM-UWB is that the power of each channel is scaled by M. This is equivalent to reducing the RF bandwidth by the same factor and hence the $\mathrm{SNR_{in}}$ will be scaled as well. Equation (3.9) can then be modified accordingly to estimate the output SNR:

$$\mathrm{SNR_{out}} = \frac{B_{RF}}{B_{SC}} \frac{\mathrm{SNR_{in}^2}/M^2}{1 + 4\mathrm{SNR_{in}}/M}. \qquad (3.31)$$

The probability of error is then calculated in the same way as for the single user case. One advantage of the proposed modification compared to the described multi-user scheme is that a larger number of channels can be used in the same bandwidth. If orthogonal sub-carrier frequencies are used, there will be no interference between the channels on the receiver side (in that sense the proposed scheme resembles OFDM). In the multi-user case, transmitters would have to be perfectly synchronized to preserve orthogonality, which is practically impossible, and as a result produces interference among different users. The only way to solve this is to separate and filter out the unwanted channels.

An existing degree of freedom in the proposed modulation technique is the sub-channels scaling. If different receivers are located at different distances from the transmitting node, the received power, and subsequently the BER, may vary. This can be circumvented by using a different scaling factor for each of the channels. Smaller scaling factor could be assigned to more distant receivers, in order to improve the BER on their sub-channels. As long as the sum off all scaling factors is 1, the maximum frequency deviation will remain the same, maintaining the signal spectrum within the defined limits.

3.3 State-of-the-Art FM-UWB Transceivers

One of the main advantages of the FM-UWB is the simplicity of the transceiver architecture, which offers a low power consumption and a high degree of integration. Different transmitter and receiver implementations have been presented in the literature. They will be discussed in the following paragraphs, with a focus on both architecture and circuit level techniques. Finally, FM-UWB will be compared to state of the art narrow-band and IR-UWB receivers, to gain insight into some of the advantages and drawbacks of the chosen modulation scheme.

3.3.1 FM-UWB Receivers

Different FM-UWB receiver architectures found in the literature are presented in Figure 3.9. The originally proposed wideband FM demodulator based on a delay line demodulator is depicted in Figure 3.9(a). Two other implementations are based on an FM discriminator, they rely on filtering to convert the input FM signal into an amplitude modulated (AM) signal. Conversion characteristics of all the demodulators are shown in Figure 3.10.

The FM-AM characteristic of the delay line demodulator was studied in the previous section (Equation (3.7)). The output AM signal will be a sine function of the input frequency. It can be seen that the choice of delay is a trade-off between the conversion gain and the bandwidth of the demodulator. Decreasing delay leads to lower conversion gain, but also increases the useful frequency range. In addition, this delay is constrained to a discrete set of values and must be equal to an odd multiple of the quarter period of the carrier frequency. It must be determined precisely in order to avoid frequency offset. In practice, a small offset will always be present as a result of process variation, however since the transmitted signal is at least 500 MHz wide, this offset should not have a major impact on the receiver performance. The first fully integrated FM-UWB receiver based on a DL demodulator was described in [11]. It achieves a sensitivity of -88 dBm while consuming 9.4 mW. The demodulator itself consumes around 5.8 mW, and the additional 3.6 mW are used by the LNA.

An LNA that provides high gain across a large bandwidth inevitably requires more power compared to a narrowband LNA. In order to reduce

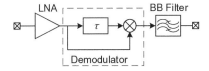

(a) Delay-Line demodulator

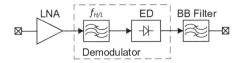

(b) Regenerative demodulator

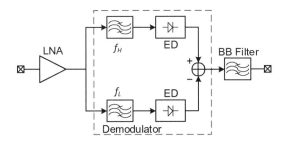

(c) Dual-Band-Pass-Filter demodulator (balanced frequency discriminator)

Figure 3.9 FM-UWB receiver architectures reported in the literature.

the power consumption, a narrow-band regenerative receiver was proposed in [12]. This approach allows for preservation of high gain and relatively good noise figure, while minimizing the power consumption. The high-Q filtering is in fact implemented in the LNA and its center frequency corresponds to either the highest or the lowest frequency of the FM-UWB signal. The band-pass filter behaves as a frequency discriminator that converts the input FM signal into an AM signal, that is then converted to IF using an envelope detector. Due to the high-Q factor of the filter that results in a very nonlinear FM-AM conversion characteristic, the demodulated signal will be a train of pulses whose frequency corresponds to the sub-carrier frequency (Figure 3.10). The receiver from [12] consumes 2.2 mW while achieving −84 dBm sensitivity. A later implementation presented in [13] introduced several improvements at the circuit level (most notably current

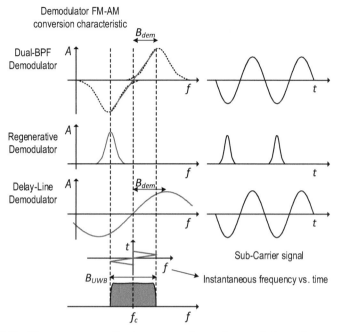

Figure   3.10   Frequency-to-amplitude   conversion   characteristic   of   reported   FM
demodulators.

reuse among several blocks) which resulted in power consumption of only
560 μW and only a slight reduction of sensitivity. Although the regenerative
receiver achieved significant power savings, there are some downsides to
this architecture. Narrow-band interferer rejection mostly relies on the high-
Q input filtering. If the interferer falls inside the pass-band it could easily
saturate the stages following the LNA and prevent reception. Indeed, such
a scenario could be avoided by introducing the on-chip tuning circuit that
could shift the filter center frequency, but this adds complexity to the system.
The second downside comes from the nonlinear FM-AM conversion. If
several FM-UWB signals were to occupy the same RF band, the weaker
signals would be attenuated in the nonlinear conversion process, which would
prevent correct demodulation. This is known as the capture effect [14], and
limits the regenerative receiver to cases where only one FM-UWB signal is
transmitted in the given RF band.

   In an attempt to improve the linearity of the regenerative demodulator,
a modified architecture was proposed in [15]. Instead of using just one
band-pass filter, a second branch was added (Figure 3.9(c)), resulting in a

Dual Band-Pass Filter (DBPF) demodulator, otherwise known as a balanced frequency discriminator. The two filters are tuned to the highest and the lowest frequency in the FM-UWB signal spectrum, they are followed by the two envelope detectors that remove the RF carrier from the signal, and the difference of the two IF signals finally yields the demodulated sub-carrier. The equivalent linearized characteristic is shown in Figure 3.10. Compared to the original regenerative receiver, the Q-factor of the two filters can be lowered, which allows some power savings per filter, but the two still consume more than the single filter from [13]. The dominant source of power consumption remains the wideband LNA, that must provide equal gain over the entire band in the DBPF receiver. The two architectures perfectly illustrate the trade-off between linearity and power consumption in FM-UWB receivers. The implementation from [15] consumed 3.8 mW, and achieved −78 dBm of sensitivity. The same architecture was reused in [10] for demodulation of a Chirp-UWB signal, where symbol-level duty-cycling of the receiver was used to bring down the average power consumption to 0.6 mW. The DBPF receiver exhibits better narrow-band interferer rejection compared to a standard regenerative receiver and should perform better in scenarios with multiple FM-UWB users, although such capability was not confirmed by measurements.

A performance summary of different FM-UWB receivers is given in Table 3.1. Each of the proposed architectures has its own advantages and disadvantages. Receiver from [11] generally has the best performance but is

Table 3.1 Performance summary of state-of-the-art FM-UWB receivers

Reference	[11]	[12, 16]	[17]	[15]	[10]	[13, 18]
Year	2009	2010	2012	2013	2014	2014
Demodulator	RF-DL	Reg	RF-DL	DBPF	DBPF	Regen.
Frequency [GHz]	7.5	3.75	3.8	3.75	8	4
Power cons. [mW]	9.4	2.2	7.2	3.8	0.6/4*	0.58
Supply [V]	1.8	1	1.6	1	1	1
Data rate [kb/s]	50	100	50	100	1000	100
Sensitivity [dBm]	−88	−84	−70	−78	−76	−80.5
NB SIR [dB]	−25	−30	–	−23	–	−18
SC-FDMA	Yes	No	–	No	No	No
Efficiency [nJ/b]	188	2.2	144	38	1	5.8
Tech. node [nm]	250	90	180	65	65	90

*Power consumption is 0.6 mW with duty-cycling and 4 mW without duty-cycling.

also the most power hungry. The regenerative receiver can provide a very low power consumption while maintaining good sensitivity, but at the cost of linearity. A trade-off between linearity and interference rejection on one side, and power consumption on the other, is demonstrated with the balanced frequency discriminator from [15]. One thing that is common for all the architectures is that the largest contributors to the power consumption are the RF blocks, mainly the LNA. Therefore, one approach to decreasing consumption would be to minimize the number of RF blocks, or to completely remove them if possible. This approach will be studied in the following chapters.

3.3.2 FM-UWB Transmitters

Unlike the FM-UWB receivers, the architecture of FM-UWB transmitters has remained unchanged over the past several years. Considering its simplicity (Figure 3.1) it is clear that there is not a lot of potential for improvement at the architectural level. In fact, the reduction of power on the transmitter side is mainly a result of improvements at the circuit level. Every FM-UWB transmitter consists of three blocks, the sub-carrier generator, the VCO (sometimes as a part of a PLL or an FLL) and a power amplifier (PA).

The sub-carrier generator synthesizes the triangular waveform that is used to drive the VCO. As the sub-carrier frequencies are rather low (typically 1–2 MHz) this block does not contribute significantly to the overall transmitter consumption. One way to implement it is a Direct Digital Synthesis (DDS) as described in [19]. The advantages of digital implementation are the simple and precise frequency control without the need for calibration. The drawback of the fully digital approach becomes apparent at higher data rates, where higher sub-carrier frequencies are needed. In [8] 51 MHz sub-carrier frequency is used. Since roughly 20 points per period are needed to generate a reliable sub-carrier waveform, a DDS would need to operate at a clock speed of more than 1 GHz, which would be difficult to implement and would consume a significant amount of power. Instead, a relaxation oscillator is used within a PLL, a simpler and lower power solution in this case. Another interesting approach that leads to a very low power consumption is a free-running relaxation oscillator that is periodically calibrated using an FLL [20]. In this case, a digital frequency control is provided through a capacitor bank, however this approach is usually not precise enough if multiple sub-carrier channels are to be used. Additionally, it might occupy a larger area due to the size of capacitors needed at the frequency of interest.

The two main parts of the FM-UWB transmitter that essentially determine its power consumption are the VCO and the PA. In cases where the transmitted power is 10 dBm or more, the transmitter efficiency is dominated by the PA, however at lower output powers, such as -10 dBm the contribution of the VCO becomes quite significant. In some of the earlier implementations, the RF carrier was synthesized using an LC VCO within a PLL [8, 21]. To decrease power, the frequency synthesizer is duty cycled, making the frequency dividers active for only 10% of the time. Although this allowed some savings, the power consumption was still on the order of 10 mW. A significant improvement was made when the LC oscillator was replaced with a ring oscillator [9, 20]. This was possible owing to the loose phase noise constraints of the FM-UWB modulation. Additionally, instead of the quasi-continuous PLL, an FLL calibration loop was used [20]. Since the FM-UWB spectrum is very wide, the center frequency can deviate slightly without a major impact on performance and it does not need to be monitored continuously. Therefore, once calibrated, the VCO can operate in a free running manner until temperature or some other external factor causes a significant frequency shift. Since these external processes are usually slow, calibration only needs to be done once in a few hours or days, which makes the average power consumption of such an FLL practically negligible. The described approach led to the first sub-milliwatt FM-UWB transmitter [20]. The next step in reducing the VCO consumption was reducing the frequency of oscillation. Since an N-stage ring oscillator produces N equally spaced phases, these phases can be combined to produce a frequency that is N times higher [22]. It is then possible to use a ring oscillator that works at a frequency that is N times lower than the carrier center frequency. The approach was demonstrated in [22] and used for the FM-UWB transmitter in [13] to reduce the power consumption down to 0.63 mW. A three-stage ring was used that oscillated at one third of the carrier frequency, which resulted in the VCO power consumption of less than 90 μW.

Even though the VCO cannot be neglected, the PA remains the most power-hungry block in the system. The key to further reducing the power consumption of an FM-UWB transmitter is an efficient power amplifier. However, design of an integrated PA for such a low power and wide band poses a number of challenges. In standard narrow-band applications targeting 10 dBm output power or more, the most efficient approach is to use a switching PA such as class D or E. The first problem with class E is that the output matching network is set to a very narrow range of frequencies and achieving good efficiency over a large band would be impossible. The second problem with switching amplifiers is that their efficiency is directly related to the on-resistance of the switch, which dictates the minimum size of the output transistor. In addition, the PA must be driven by a square wave with very sharp transitions to minimize the turn-on time of the switch. The two requirements impose very hard constraints, resulting in power dissipation in the driving circuits that is comparable to that of the PA. Therefore, when the driving circuit is also accounted for, switching PAs seem not to be the best solution. The linear power amplifiers, classes A, AB, B and C, do not achieve as high efficiency, but their driving requirements are also lower. Moving from class A to class C operation, the maximum attainable efficiency increases, but the power gain decreases and larger driving signal is necessary, thus again shifting the burden from the PA to the driver. A good compromise is the class AB that attains decent efficiency and does not need a rail-to-rail input signal. In fact, all of the transmitters reported in [9, 13, 20], which achieve the lowest power consumption reported so far, use a complementary class AB power amplifier.

For linear PAs in general, optimal efficiency is obtained when the output voltage swing is maximized. In case of a complementary class AB or B amplifier, maximum output voltage swing is equal to the supply voltage. The load resistance seen from the power amplifier, must therefore be chosen such as to provide the desired output power. The problem with low power transmission is that the value of the optimal load resistance is relatively high. As a consequence a large transformation ratio of the matching network is needed, which then increases the losses in the network. One way to solve this problem would be to reduce the supply voltage. However, such an approach would require another circuit that would lower the voltage to the desired level, such as a DC-DC converter. This would not only increase complexity but also introduce its own losses and possibly require off-chip components. A better and simpler way is to apply current reuse technique demonstrated in [13], where the PA and the driver share the same current. Since the effective PA supply voltage is lower, there is no need for such a high transformation ratio of the matching network and, at the same time, the PA bias current is used to supply the driver. The efficiency of the transmitter can therefore be improved without any increase in complexity, which led to the current lowest power FM-UWB transmitter [13, 18], as shown in Table 3.2.

3.3.3 FM-UWB Against IR-UWB and Narrowband Receivers

So far the main characteristics of FM-UWB have been discussed, the potential of FM-WUB has been shown and the existing implementations of FM-UWB

Table 3.2 Performance summary of state-of-the-art FM-UWB transmitters

Reference	[21]	[8]	[20]	[17]	[9]	[10]	[13, 18]
Year	2010	2011	2011	2012	2013	2014	2015
SC Modulation	2-FSK	2-FSK	2-FSK	2-FSK	8-FSK	2-FSK	2-FSK
Frequency [GHz]	3.8	3.8	4	3.8	3.75	8	4
Bandwidth [MHz]	600	700	500	560	500	500	500
Power cons. [mW]	9.6	18.2*	0.9	8.7	1.14	3.5**	0.63
Supply [V]	1.6	1.6	1	1.6	1	1	1
Data rate [kb/s]	10	1000	100	50	750	1000	100
Out. power [dBm]	−14.5***	−12.8	−10.2	−13.7	−14	−11***	−10.1
Efficiency [nJ/b]	960	18.2*	9	174	1.5	0.39	3.1
Tech. node [nm]	180	180	90	180	65	65	90

*Excluding the output PA.
**In continuous mode, 0.39 mW with duty cycling.
***Estimated from figure.

transmitters and receivers have been presented. The question now is how does the FM-UWB compare to other low power modulation schemes? In Figure 3.11 FM-UWB receivers are placed together with low power receivers from Chapter 2, showing the data rate against power consumption. They consume a somewhat lower power than BLE receivers, but also target lower speed. In terms of power consumption they cannot achieve nanowatt levels of wake-up receivers.

As explained previously, if FM-UWB is compared to a standard FSK modulation, there is an inherent loss in sensitivity. Unfortunately, this is an unavoidable drawback of FM-UWB. If a narrowband receiver and an FM-UWB receiver using the same data rate perform similarly in terms of noise figure, the narrowband receiver will provide better sensitivity. This can be observed in Figure 3.12, where efficiency of receivers is plotted against sensitivity. The FM-UWB receivers cannot achieve the same sensitivity at comparable efficiency levels as the narrowband receivers. However, FM-UWB provides other benefits that may not be apparent at first. It is inherently robust against interferers, unlike NB radios that need to rely on filtering. Owing to the spread spectrum, FM-UWB is also robust against frequency selective fading. Narrowband radios might be unable to establish a link due to a notch in the channel frequency characteristic, whereas the FM-UWB only suffers a minor performance degradation. Also, FM-UWB

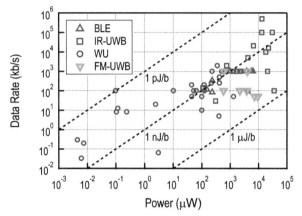

Figure 3.11 Comparison of FM-UWB receivers and other low power receivers from Chapter 2, data-rate against power consumption.

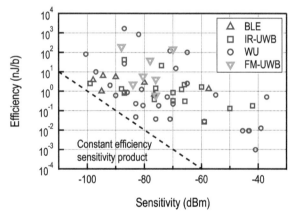

Figure 3.12 Comparison of FM-UWB receivers and other low power receivers from Chapter 2, efficiency against sensitivity.

could provide support for multi-user communication at almost no increase in power consumption. Finally, FM-UWB has better potential for minia-turization, enabling implementations with no off-chip components. Every narrowband radio needs a crystal oscillator to provide a precise frequency reference, and in most cases other off-chip components are needed to provide additional filtering, or output matching. Thanks to robustness to reference frequency offset, that partially comes from the large signal bandwidth, FM-UWB is capable of using an imprecise, on-chip reference oscillator, while still providing reliable communication. The combination of robustness, architecture simplicity and high degree of integration are, finally, the main arguments in favor of FM-UWB when compared to narrowband radios.

IR-UWB receivers cover a very wide range of data rates, from kb/s to almost Gb/s, while maintaining an almost constant efficiency. This is possible assuming symbol-level duty cycling can be applied. In order to benefit from symbol-level duty cycling the IR-UWB receivers must have a good timing reference, and require initial synchronization, both of which add complexity and cost to the system. The FM-UWB generally requires a fairly simple receiver architecture and has a lower peak power consumption making it cheaper and more appealing for battery powered systems. One other advantage of FM-UWB compared to IR-UWB is the multi-user capability. FM-UWB devices can transmit in the same RF band at the same time. A similar TDMA based scheme at the symbol level would be possible with IR-UWB wherein each transmitter has a time slot in which it can transmit a pulse during one symbol period. However, this would require a nanosecond level synchronization among the nodes, adding a prohibitively high level of complexity to the system.

3.4 Summary

The first part of this above chapter describes the main principles of the FM-UWB modulation. Basic calculations related to the modulation technique are presented and extended to the cases with multiple users. The described techniques, such as multi-user communication and multi-channel transmission, can be used to optimize the system performance according to the specific needs. Different sub-channels can be used, trading data-rate per channel with the number of available sub-channels, depending on the number of nodes in the network and their purpose.

In the second part of the chapter, the state of the art FM-UWB transceivers are discussed along with the the most important power reduction techniques reported in the literature. These techniques, combined with technology scaling, led to sub-milliwatt power consumption levels in today's implementations. The evolution of power consumption over the past 8 years is illustrated in Figure 3.13 for both transmitters and receivers, from which a decrease by a factor of 20 can be observed. However, the narrow-band receivers still have the edge, at least with respect to power consumption. The proposed wake-up receivers found in literature consume from $100\,\mu$W [23] all the way down to $100\,$nW [24]. FM-UWB can hardly compete with such low

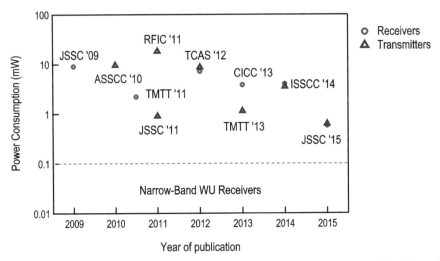

Figure 3.13   FM-UWB transmitters and receivers, evolution of power consumption. Type of demodulator used in each receiver is indicated on the graph.

levels, a simple consequence of the fact that wide-band circuits require more power to achieve the same performance in terms of gain and noise figure. On the other hand, the FM-UWB brings higher resilience to interferers, without off-chip components such as SAW filters, better performance in frequency selective channels and higher potential for miniaturization. All of these are very favorable capabilities that could assure a place for FM-UWB in short-range applications such as wireless body area networks.

References

[1] J. F. M. Gerrits, M. H. L. Kouwenhoven, P. R. van der Meer, J. R. Farserotu, and J. R. Long, "Principles and Limitations of Ultra-wideband FM Communications Systems," *EURASIP J. Appl. Signal Process.*, vol. 2005, pp. 382–396, Jan. 2005.

[2] M. Kouwenhoven, *High-Performance Frequency-Demodulation Systems*. Delft University Press, 1998.

[3] N. Saputra and J. Long, *FM-UWB Transceivers for Autonomous Wireless Systems:*, ser. River Publishers Series in Circuits and Systems. River Publishers, 2017.

[4] J. F. M. Gerrits, J. Farserotu, and J. Long, "Multipath Behavior of FM-UWB Signals," in *IEEE International Conference on Ultra-Wideband, 2007. ICUWB 2007*, Sep. 2007, pp. 162–167.

[5] ——, "Multi-user capabilities of UWBFM communications systems," in *IEEE International Conference on Ultra-Wideband*, Sep. 2005, pp. 1–6.

[6] V. Kopta, J. Farserotu, and C. Enz, "FM-UWB: Towards a robust, low-power radio for body area networks," *Sensors*, vol. 17, no. 5, 2017.

[7] *IEEE Standard for Local and Metropolitan Area Networks – Part 15.6: Body Area Networks*, 2012.

[8] B. Zhou, H. Lv, M. Wang, J. Liu, W. Rhee, Y. Li, D. Kim, and Z. Wang, "A 1 mb/s 3.2–4.4 ghz reconfigurable FM-UWB transmitter in 0.18 μm CMOS," in *2011 IEEE Radio Frequency Integrated Circuits Symposium (RFIC)*, June 2011, pp. 1–4.

[9] F. Chen, Y. Li, D. Lin, H. Zhuo, W. Rhee, J. Kim, D. Kim, and Z. Wang, "A 1.14 mw 750 kb/s FM-UWB transmitter with 8-FSK subcarrier modulation," in *2013 IEEE Custom Integrated Circuits Conference (CICC)*, Sep. 2013, pp. 1–4.

[10] F. Chen, Y. Li, D. Liu, W. Rhee, J. Kim, D. Kim, and Z. Wang, "9.3 A 1 mw 1 mb/s 7.75-to-8.25 ghz chirp-UWB transceiver with low peak-power transmission and fast synchronization capability," in *Solid-State Circuits Conference Digest of Technical Papers (ISSCC), 2014 IEEE International*, Feb. 2014, pp. 162–163.

[11] Y. Zhao, Y. Dong, J. F. M. Gerrits, G. van Veenendaal, J. Long, and J. Farserotu, "A Short Range, Low Data Rate, 7.2 GHz-7.7 GHz FM-UWB Receiver Front-End," *IEEE Journal of Solid-State Circuits*, vol. 44, no. 7, pp. 1872–1882, July 2009.

[12] N. Saputra and J. Long, "A Short-Range Low Data-Rate Regenerative FM-UWB Receiver," *IEEE Transactions on Microwave Theory and Techniques*, vol. 59, no. 4, pp. 1131–1140, Apr. 2011.

[13] N. Saputra and J. R. Long, "A Fully Integrated Wideband FM Transceiver for Low Data Rate Autonomous Systems," *IEEE Journal of Solid-State Circuits*, vol. 50, no. 5, pp. 1165–1175, May 2015.

[14] K. Leentvaar and J. Flint, "The capture effect in FM receivers," *IEEE Transactions on Communications*, vol. 24, no. 5, pp. 531–539, May 1976.

[15] F. Chen, W. Zhang, W. Rhee, J. Kim, D. Kim, and Z. Wang, "A 3.8-mW 3.5-4-GHz Regenerative FM-UWB Receiver With Enhanced Linearity by Utilizing a Wideband LNA and Dual Bandpass Filters,"

IEEE Transactions on Microwave Theory and Techniques, vol. 61, no. 9, pp. 3350–3359, Sep. 2013.

[16] N. Saputra, J. R. Long, and J. J. Pekarik, "A 2.2 mW regenerative FM-UWB receiver in 65 nm CMOS," in *2010 IEEE Radio Frequency Integrated Circuits Symposium*, May 2010, pp. 193–196.

[17] B. Zhou, J. Qiao, R. He, J. Liu, W. Zhang, H. Lv, W. Rhee, Y. Li, and Z. Wang, "A Gated FM-UWB System With Data-Driven Front-End Power Control," *IEEE Transactions on Circuits and Systems I: Regular Papers*, vol. 59, no. 6, pp. 1348–1358, June 2012.

[18] N. Saputra, J. Long, and J. Pekarik, "A low-power digitally controlled wideband FM transceiver," in *2014 IEEE Radio Frequency Integrated Circuits Symposium*, June 2014, pp. 21–24.

[19] P. Nilsson, J. F. M. Gerrits, and J. Yuan, "A Low Complexity DDS IC for FM-UWB Applications," in *2007 16th IST Mobile and Wireless Communications Summit*, July 2007, pp. 1–5.

[20] N. Saputra and J. Long, "A Fully-Integrated, Short-Range, Low Data Rate FM-UWB Transmitter in 90 nm CMOS," *IEEE Journal of Solid-State Circuits*, vol. 46, no. 7, pp. 1627–1635, July 2011.

[21] B. Zhou, R. He, J. Qiao, J. Liu, W. Rhee, and Z. Wang, "A low data rate FM-UWB transmitter with-based sub-carrier modulation and quasi-continuous frequency-locked loop," in *2010 IEEE Asian Solid-State Circuits Conference*, Nov. 2010, pp. 1–4.

[22] J. Pandey and B. P. Otis, "A sub-100 μW MICS/ISM band transmitter based on injection-locking and frequency multiplication," *IEEE Journal of Solid-State Circuits*, vol. 46, no. 5, pp. 1049–1058, May 2011.

[23] C. Salazar, A. Cathelin, A. Kaiser, and J. Rabaey, "A 2.4 ghz interferer-resilient wake-up receiver using a dual-if multi-stage n-path architecture," *IEEE Journal of Solid-State Circuits*, vol. 51, no. 9, pp. 2091–2105, Sep. 2016.

[24] N. E. Roberts and D. D. Wentzloff, "A 98 nw wake-up radio for wireless body area networks," in *2012 IEEE Radio Frequency Integrated Circuits Symposium*, June 2012, pp. 373–376.

4

The Approximate Zero IF Receiver Architecture

4.1 Introduction

One of the primary goals of this work is to reduce the power consumption of an FM-UWB transceiver. In duty cycled wireless sensor networks the bottleneck is typically the receiver. This is because transmitters only need to be turned on when there is a need to transmit data. As a result their power consumption will only be a small fraction of the overall power consumed by the network. Receivers, on the other hand, need to capture the transmitted data and must therefore be turned on periodically to check whether data is being transmitted. This is why the power consumption of the network will almost entirely be determined by the receiver power consumption. Lowering the duty cycle ratio, or equivalently, increasing the duration of the period between the two on states of the receiver, can be used to bring down network power consumption, but it will also increase latency.

In body area networks, such as [10], where sensors need to provide pressure information from the prosthetic limb to the patient, latency constraints are imposed by the physiological characteristics of the human body. In order to provide a natural sense of touch, the delay from sensors to actuators must not be larger than the time it takes for neurons to convey information from the fingers to the brain. Once the maximum delay limit is reached the only way to reduce network power consumption is to reduce the consumption of the FM-UWB receiver.

Another property that could be of use in the receiver is the capability to handle multiple FM-UWB signals at the same time. This requirement comes from the fact that potentially a large number of sensor nodes may be located close to each other. Providing the multi-user capability would then allow to parallelize data transfer and decrease network delay. Normally, receivers with such capability need good linearity and dynamic range, which again come at the price of power consumption. If the distance between nodes is not large, and WBAN is a typical example of such an application, sensitivity is not a limiting factor, and can be sacrificed for the benefit of multi-user communication and energy efficiency.

This chapter describes the proposed architecture, intended to further reduce the power consumption of an FM-UWB receiver. Two variations of the architecture are explored, one that aims to provide the multi-user communication capability, and another that attempts to aggressively lower the power consumption, while essentially neglecting all other aspects.

4.2 The Uncertain IF Architecture

It is a common observation that in any type of circuits there is a correlation between the frequency of operation and power consumption. The simplest example is the CMOS logic gate, where it can be shown that the dynamic power consumption is proportional to the frequency of operation $P_{dyn} \propto fCV^2$. Similar conclusion holds for other types of circuits, for example, amplifiers typically need more power to achieve the same gain at higher frequencies (or to provide larger bandwidth). As a consequence, the most power-hungry blocks in the receivers are the ones that operate at RF. These are usually the Low Noise Amplifier (LNA), needed to provide a good noise figure, and the frequency synthesizer, where the dominant consumers are the voltage controlled oscillator and frequency dividers. A typical example could be the Bluetooth receiver presented in [1], where the LNA consumes around 25% of the overall power, and 53% of the power is used for the PLL, including the DCO.

Due to the large bandwidth of the FM-UWB signal, precise frequency synthesizers can be completely removed. Gain at high frequencies, however, remains a bottleneck. Consider the FM-UWB receiver from [2], where the LNA consumes 1.6 mW, or 73% of the overall power consumption. A similar case is found in [3], where the LNA consumes 55% of the entire receiver consumption. The preamplifier and the demodulator in the FM-UWB receiver from [4], both operating at RF, consume around 3 mA and 6 mA, respectively. Removing the LNA from the design, or loosening the specifications on RF gain and noise figure, could lead to significant power savings.

The opportunity to decrease power consumption by moving the gain stages from RF to IF was first recognized by Pletcher [5], who demonstrated this approach through the implementation of the "Uncertain IF" receiver. In this design the LNA is merged with the mixer into a single current reuse

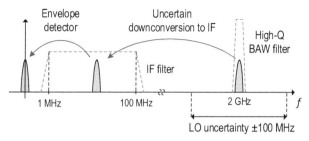

Figure 4.1 Principle of operation of the uncertain IF receiver.

block, the active mixer. Combined with the external bulk acoustic wave (BAW) resonator, it provides the input matching, and converts the RF signal to IF, where the gain stages are located. The lack of voltage gain at RF simplifies the design and allows for low dc current of the active mixer. Although the power hungry LNA is eliminated from the design, the LO signal must now be generated in order to perform the downconversion. For the proposed design from [5] to be truly power efficient, the LO generator must consume sufficiently low power. A simple three-stage, CMOS ring oscillator used in this design consumes very little power and provides a rail to rail output signal, but this comes at a price. Ring oscillators are generally sensitive to changes in supply voltage and temperature, phase noise is relatively high compared to LC oscillators, and finally the output frequency is rather unstable and tends to drift with time. This is the key point of the architecture from [5]. Instead of using a PLL to stabilize the oscillator frequency, the IF amplifier bandwidth is increased to allow for LO frequency offset. Hence the name "uncertain IF" receiver is used. The described principle is illustrated in Figure 4.1. The designed oscillator generates the LO signal that remains within the ±100 MHz range from the center frequency. To account for this offset, the IF amplifier is designed with a bandwidth of 100 MHz, assuring that the downconverted signal falls inside the desired band. Using 100 MHz amplifiers to amplify a signal with 10 kHz bandwidth is a necessary overhead, but still results in less power consumed than if a PLL were used to generate the LO signal. The oscillator, however, does need to be calibrated periodically to compensate for the drift due to temperature or supply voltage variation, and maintain the LO frequency within the defined limits. Owing to the fact that these changes are slow, the calibration will only be done once in a few hours, resulting in a negligible overhead in terms of power consunmption. The final stage of the receiver is the envelope detector that demodulates the transmitted OOK (on-off keying) signal. It should still be noted that a relatively good noise performance in this case was achieved using a narrowband BAW filter that is precisely tuned to the frequency of the transmited signal. Finally, the receiver from [5] reached a power consumption of only $50\,\mu$W, for an input signal at $2\,$GHz, and is still among the lowest consuming narrowband receivers in the literature today.

The same principle can be applied to the FM-UWB signal. Instead of implementing the wideband amplifiers at RF, the input signal is directly downconverted to zero center frequency using an active mixer (LNA and mixer stack), allowing amplification and processing to be done at low frequencies. Since the LO signal is generated using an imprecise ring oscillator, a certain frequency offset will always be present between the LO signal and the center frequency of the input signal. Hence, the proposed receiver is referred to as the "Approximate Zero IF" receiver. The aforementioned offset should be relatively small compared to the $500\,$MHz wide input signal, and so the necessary overhead, i.e. larger bandwidth of the IF amplifiers and the demodulator, will be relatively small. Moving the main gain stages from RF to IF results in higher noise figure (NF) of the receiver chain. Since only a limited amount of gain is available at high frequencies, noise of the IF stages will contribute more significantly to the overall noise figure. At the same time, the power consumed by the IF amplifiers can now be greatly reduced since they operate at low frequencies instead of RF, and the same overall gain comes at a lower price.

4.3 The Approximate Zero IF Receiver with Quadrature Downconversion

The first proposed receiver architecture is based on a delay line demodulator described in the previous chapter. This demodulator has already been used in some receiver implementations [4, 6]. In its original form it operates directly at RF, and the delay needs to be tuned precisely to the signal center frequency. The demodulator can be moved from RF to baseband, without changing its functionality, if two signal branches are used with a 90° phase shift between them [7]. The two signals can easily be generated by using quadrature LO signals for downconversion.

The proposed receiver architecture is shown in Figure 4.2. Some of the blocks that will be present in the actual implementation are omitted for clarity, and to emphasize functionality. The input signal is directly converted to zero frequency by the mixer. Since a ring oscillator will be used to generate

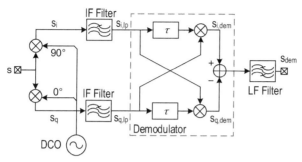

Figure 4.2 Block diagram of approximate zero IF receiver with IQ downconversion.

the LO signal, some frequency offset will always be present, meaning that the downconverted signal will never be precisely centered at zero, and the architecture is therefore named the "Approximate Zero IF" architecture. The downconverted signal will then be amplified by the IF amplifiers. In Figure 4.2 only filters are shown to emphasize limited bandwidth of the IF path. In this work a 500 MHz wide FM-UWB signal is used. After downconversion, the signal would ideally occupy frequencies from zero to 250 MHz. Accounting for a frequency offset of ±50 MHz, 300 MHz should be sufficient for the bandwidth of the IF path. The case remains the same with the demodulator bandwidth. It needs to be larger in order to accommodate the input signal with a frequency offset.

The principle of operation of the demodulator can be described through the following mathematical model. The simple calculation presented here follows the approach from [8], extending it to account for the LO frequency offset. The aim of the calculation is to explain the principle of operation and provide insights into the main trade-offs when choosing the delay of the demodulator, which is the main design parameter in this case and determines the bandwidth of the demodulator. The FM-UWB signal at the input of the receiver can be represented as:

$$s(t) = A\cos(\omega_c t + \phi(t)), \tag{4.1}$$

where $\omega_c = 2\pi f_c$ is the carrier center frequency of the signal and $\phi(t)$ is the time varying phase, which is the integral of the sub-carrier wave (usually a periodic triangular or sine signal):

$$\phi(t) = \Delta\omega \int_{-\infty}^{t} m(t)\mathrm{d}t. \tag{4.2}$$

The sub-carrier wave $m(t)$ is normalized to the interval $[-1,1]$, and $\Delta\omega$ is the frequency deviation corresponding to half of the FM-UWB signal bandwidth $\Delta\omega = 2\pi\Delta f = \pi B_{UWB}$. The signal is first converted to baseband, such that the signals in the in-phase and quadrature branches are given by

$$s_i(t) = A\cos(\omega_c t + \phi(t))\cos(\omega_{osc}t) \tag{4.3}$$
$$s_q(t) = A\cos(\omega_c t + \phi(t))\sin(\omega_{osc}t) \tag{4.4}$$

where the conversion gain of the mixer is assumed to be unity for simplicity. Note that since the carrier frequency ω_c does not correspond ideally to the locally generated frequency ω_{osc}, there will be a residual term after mixing that is equal to the difference of the two frequencies. The two filtered, downconverted components at the demodulator input are:

$$s_{i,lp}(t) = \frac{A}{2}\cos(\omega_{off}t + \phi(t)) \tag{4.5}$$

$$s_{q,lp}(t) = -\frac{A}{2}\sin(\omega_{off}t + \phi(t)) \tag{4.6}$$

where $\omega_{off} = 2\pi f_{off} = \omega_c - \omega_{osc}$ is the frequency offset of the LO signal. The two quadrature signals are then multiplied with the delayed copy of each other in the process of demodulation. The signals at the output of the two demodulator mixers are:

$$s_{i,dem}(t) = -\frac{A^2}{4}\cos(\omega_{off}(t-\tau) + \phi(t-\tau))\ \sin(\omega_{off}t + \phi(t)) \tag{4.7}$$
$$s_{q,dem}(t) = -\frac{A^2}{4}\cos(\omega_{off}t + \phi(t))\ \sin(\omega_{off}(t-\tau) + \phi(t-\tau)). \tag{4.8}$$

Finally, the difference of $s_{i,dem}(t)$ and $s_{q,dem}(t)$ results in the following signal:

$$s_{dem}(t) = \frac{A^2}{4}\sin(\omega_{off}\tau + \phi(t) - \phi(t-\tau)). \tag{4.9}$$

Following the same approach as in [8], assuming the time interval τ is small enough that $\phi(t)$ does not change too significantly, Equation (4.9) can then be approximated by

$$s_{dem}(t) \approx \frac{A^2}{4}\sin(\omega_{off}\tau + \tau\frac{d\phi(t)}{dt}) \tag{4.10}$$

$$= \frac{A^2}{4}\sin(\omega_{off}\tau + \tau\Delta\omega\, m(t)), \tag{4.11}$$

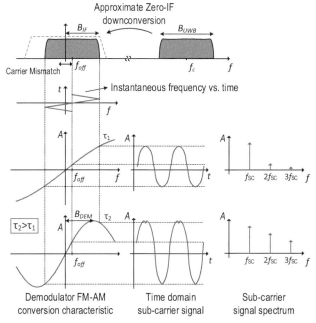

Figure 4.3 Principle of operation of approximate zero IF receiver with IQ downconversion.

which in fact corresponds to the demodulated signal. The last equation reveals the sinusoidal FM-AM characteristic of the demodulator, showing that it, in fact, acts as a baseband equivalent of the RF delay line demodulator. The illustration of the demodulation principle and the equivalent demodulator characteristic are given in Figure 4.3. The shape of the output demodulated signal is shown for two different values of the delay, assuming that a certain frequency offset is present in the LO signal. The amplitude and the shape of the demodulated signal depend on the demodulator delay τ and the frequency offset f_{off}, as illustrated in Figure 4.3. The demodulator bandwidth can be defined as the monotonic part of the characteristic, i.e. $\tau \times B_{DEM} = \pi/2$. Increasing the delay results in decreased demodulator bandwidth, which increases the amplitude, but also distorts the output signal. Note that unlike with the RF delay line demodulator, the delay τ is no longer related to the input signal center frequency. Ideally, in the case of the RF delay line demodulator, τ should be equal to the integer multiple of the quarter period of the center frequency $NT/4$. Any deviation of τ results in mismatch between the demodulator center frequency and the signal center frequency.

The end effect is equivalent to the LO frequency offset in the approximate zero IF receiver.

The conversion gain is defined as the ratio of the fundamental amplitude of the demodulated signal and the amplitude of the signal at the demodulator input:

$$G_{conv} = \frac{A_{(1)}}{A/2} = \frac{C_1 A^2/4}{A/2} = \frac{A}{2}C_1. \tag{4.12}$$

The coefficient C_1 corresponds to the fundamental component of the demodulated signal normalized to $A^2/4$ and accounts for the non-linear characteristic of the demodulator. This coefficient will depend on τ and f_{off}. Assuming a triangular sub-carrier wave, C_1 and C_2 (normalized second harmonic amplitude) are calculated and plotted in Figure 4.4 as functions of the frequency offset, for several different values of the delay τ. These graphs show the trade-off between the distortion and the conversion gain mentioned above. Choosing larger τ, such that the demodulator bandwidth is smaller than the signal bandwidth, for example $B_{dem} = 200\,\text{MHz}$, will indeed result in a higher gain, but will also make it more sensitive to the carrier offset. For the value of τ selected to provide 500 MHz bandwidth, the conversion gain is practically half of that obtained for $B_{DEM} = 200\,\text{MHz}$, but remains almost constant even for a very high carrier offset. The conversion gain remains the same for both noise and signal at the input, and in that sense doesn't affect the SNR. However, in a practical realization the demodulator itself will generate noise, and with this noise taken into account higher conversion gain will yield a higher output SNR. It should also be noted that a high demodulator bandwidth (together with IF bandwidth B_{IF}) also results in a higher noise bandwidth, which combined with lower gain inevitably leads to a degradation of sensitivity. Looking at the second harmonic, the increase of distortion that comes with the decrease of bandwidth becomes evident. In the ideal case, with no carrier offset, the second harmonic will be zero. However, for the proposed receiver architecture this will never be the case and the amplitude of the second harmonic will depend on offset and demodulator delay. Finally as a compromise between the gain, sensitivity to frequency offset and distortion, a bandwidth of 300 MHz can be chosen for the demodulator implementation and the same value should be used for the bandwidth of the preceding IF amplifiers.

The proposed demodulator can be used to simultaneously demodulate two or more FM-UWB signals. If an additional FM-UWB signal, occupying the

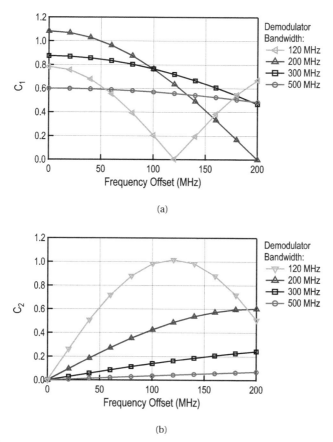

Figure 4.4 Normalized fundamental C_1 and second harmonic amplitude C_2 at the output of the demodulator vs. the offset frequency. First harmonic is proportional to conversion gain. Four curves are plotted for four different values of the demodulator bandwidth (or equivalently different values of the delay τ).

same RF bandwidth but using a different sub-carrier frequency, is present at the input of the receiver, the signals in the I and Q branches are given by

$$s_{i,lp} = \frac{A_1}{2}\cos(\omega_{off}\tau + \phi_1(t)) + \frac{A_2}{2}\cos(\omega_{off}\tau + \phi_2(t)) \qquad (4.13)$$

$$s_{q,lp} = -\frac{A_1}{2}\sin(\omega_{off}\tau + \phi_1(t)) - \frac{A_2}{2}\sin(\omega_{off}\tau + \phi_2(t)). \qquad (4.14)$$

Following the same steps as in the above calculation the demodulated signal can then be derived as

$$s_{dem} = \frac{A_1^2}{4} \sin(\omega_{off}\tau + \tau\frac{d\phi_1(t)}{dt}) + \frac{A_2^2}{4} \sin(\omega_{off}\tau + \tau\frac{d\phi_2(t)}{dt}) + W(t).$$
(4.15)

Aside from the first two terms, which represent the two demodulated signals, an additional term, $W(t)$ appears. This term corresponds to the intermodulation product of the two FM-UWB signals. The effect is the same as for the case of the RF delay line demodulator, where the additional term corrupts the two useful signals and limits the achievable BER. Fortunately, as will be shown, this term will be spread across a large frequency, allowing to filter out most of it in the baseband. The $W(t)$ term is given by

$$W(t) = \frac{A_1 A_2}{4} \sin(\omega_{off}\tau + \phi_1(t) - \phi_2(t - \tau))$$
$$+ \frac{A_1 A_2}{4} \sin(\omega_{off}\tau + \phi_2(t) - \phi_1(t - \tau)).$$
(4.16)

Using the same approximation as in the single-user case, assuming τ is very small $W(t)$ can be rewritten as

$$W(t) \approx \frac{A_1 A_2}{4} \sin\left(\omega_{off}\tau + \frac{\tau}{2}\frac{d\phi_1(t)}{dt} + \frac{\tau}{2}\frac{d\phi_2(t)}{dt}\right) \sin\left(\phi_1(t) + \phi_2(t)\right)$$
(4.17)

$$= \frac{A_1 A_2}{2} \sin\left(\omega_{off}\tau + \frac{\tau\Delta\omega}{2}(m_1(t) + m_2(t))\right)$$
$$\times \sin\left(\Delta\omega \int_{-\infty}^{t} (m_1(t) - m_2(t))dt\right)$$
(4.18)

$$= \frac{A_1 A_2}{4} w(t).$$
(4.19)

The intermodulation product $W(t)$ consists of two factors. The first one, proportional to the sum of the two sub-carrier signals, is the slow-varying envelope. Clearly the shape of the envelope will depend on the demodulator delay τ, and the frequency offset f_{off}, and therefore these two parameters will affect the average power of the intermodulation product. The second factor is spread from 0 to $2\Delta\omega$, which is equal to the signal bandwidth B_{UWB}, with the instantaneous frequency that is proportional to the difference of the two sub-carrier signals. Since the intermodulation product is spread over a very wide band, only a small fraction of its power will fall into the useful sub-carrier band B_{SC}. The effect of inter-user interference will be similar to the elevated noise floor at the output of the receiver. This results in a degradation of sensitivity as either the number of users or the power of additional users increase. For a given targeted bit error rate, interference among users will ultimately limit the number of users or the difference in power levels between the two FM-UWB signals that can be handled at the same time.

The average power of the intermodulation product can be calculated as

$$\overline{W(t)^2} = \frac{A_1^2 A_2^2}{16}\overline{w(t)^2} = \frac{A_1^2 A_2^2}{16}C_{MU} \tag{4.20}$$

Factor C_{MU} is the normalized average power of the intermodulation product and depends on τ and f_{off}. It is calculated for two triangular sub-carrier waves and presented in Figure 4.5(a). The decrease of the demodulator bandwidth (increase of τ) in this case leads to increased power of the inter-modulation product. Making the approximation that the spectrum of $W(t)$ is flat across the entire band [8], the output signal-to-interference ratio (SIR) can be calculated as

$$SIR_{out} = 10\log_{10}\left(\frac{A_1^2|C_1|^2}{A_2^2 C_{MU}}\frac{B_{UWB}}{B_{SC}}\right), \tag{4.21}$$

where factor $|C_1|^2/C_{MU}$ is added to the original formula from [8] to account for the frequency offset [9]. Figure 4.5(b) shows how this factor changes with the frequency offset for different demodulator bandwidths. As expected the best result is obtained for the highest demodulator bandwidth. The difference is, however, not too significant compared to the case with 300 MHz bandwidth. At the same time, extending the demodulator bandwidth also requires the extension of the IF amplifier bandwidth, which finally leads to increased power consumption. For this reason 300 MHz is chosen as a good trade-off between power and distortion and is the bandwidth that will be used in the receiver implementation described in the following chapter.

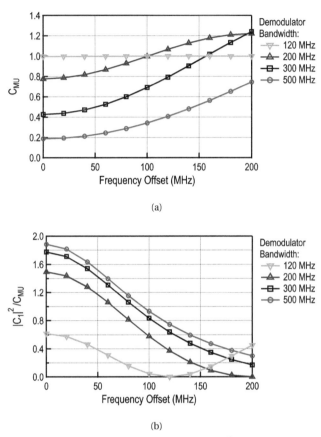

(a)

(b)

Figure 4.5   Coefficient C_{MU} (a) and correction factor $|C_1|^2/C_{MU}$ for SIR (b) as functions of the frequency offset. Four curves are correspond to three different values of the demodulator bandwidth (or equivalently values of the delay τ).

4.4 The Approximate Zero IF Receiver with Single-Ended Downconversion

As a general rule, quadrature downconversion is needed in direct downconversion receivers, otherwise part of the information will be lost, and it will be impossible to recover the data. However, because of the properties of the FM-UWB signal, transmitted bits can be recovered even if the signal is directly converted to zero using only a single branch. Shift to a single-ended receiver architecture enables some power savings. First of all, only one IF amplifier can be used, allowing to halve the power of the IF stages. In addition, the

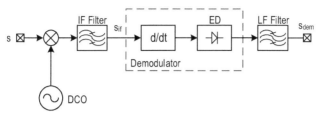

Figure 4.6 Block diagram of approximate zero IF receiver with single-ended downconversion.

simplified FM demodulator should also allow some savings compared to the IQ delay line demodulator. Finally, the most important savings come from the DCO. Quadrature demodulation requires quadrature LO generation that tends to be power costly. Using a single-ended oscillator simplifies the circuit and allows to reduce power by a factor of more than 2 for the same oscillation frequency.

The operation of the demodulator can be explained using a simplified receiver model, presented in Figure 4.6. Like in the previous case the input FM-UWB signal can be represented as

$$s(t) = A\cos(\omega_c t + \phi(t)), \tag{4.22}$$

where $\phi(t)$ is again the integral of the sub-carrier wave, and ω_c is the center frequency. After downconversion, the signal at the mixer output is given by

$$s_{mix}(t) = kA\cos(\omega_c t + \phi(t) + \phi_0)\cos(\omega_{osc}t) \tag{4.23}$$

The IF low-pass filter removes all the high frequency components, resulting in the signal at the filter output given by

$$s_{if}(t) = \frac{A}{2}\cos(\omega_{off}t + \phi(t)), \tag{4.24}$$

where ω_{off} is the offset frequency, that is equal to the difference of the LO frequency and the signal center frequency. The following stage, a differentiator, converts the FM signal into an AM signal given by

$$\frac{ds_{if}(t)}{dt} = \frac{A}{2}\sin(\omega_{off}t + \phi(t))\left(\omega_{off}\tau_0 + \tau_0\frac{d\phi(t)}{dt}\right), \tag{4.25}$$

where τ_0 is the time constant of the differentiator. The resulting signal is then demodulated using the envelope detector. Here, an ideal square law envelope detector is assumed, resulting in the output signal given by

$$
s_{dem}(t) = \frac{A^2}{4} \sin(\omega_{off}t + \phi(t))^2 \left(\omega_{off}\tau_0 + \tau_0\frac{d\phi(t)}{dt}\right)^2
$$

$$
= \frac{A^2}{8}(1 - \cos(2\omega_{off}t + 2\phi(t))) \left(\omega_{off}\tau_0 + \tau_0\frac{d\phi(t)}{dt}\right)^2 \quad (4.26)
$$

The low-pass filter following the envelope detector will practically remove the fast changing component $\cos(2\omega_{off}t + 2\phi(t))$, resulting in the signal at the filter output given by

$$
s_{dem}(t) = \frac{A^2}{8} \left(\omega_{off}\tau_0 + \tau_0\frac{d\phi(t)}{dt}\right)^2 . \quad (4.27)
$$

For simplicity, let us assume that the sub-carrier signal is a sine wave. The demodulated signal is then

$$
s_{dem}(t) = \frac{A^2}{8} (\omega_{off}\tau_0 + \Delta\omega\tau_0 \sin(\omega_{sc}t))^2 \quad (4.28)
$$

$$
= \frac{A^2}{8}\tau_0^2 (\omega_{off}^2 + 2\omega_{off}\Delta\omega \sin(\omega_{sc}t) + \Delta\omega^2 \sin^2(\omega_{sc}t)) \quad (4.29)
$$

$$
= \frac{A^2}{8}\tau_0^2 \left(\omega_{off}^2 + 2\omega_{off}\Delta\omega \sin(\omega_{sc}t)\right.
$$

$$
\left. +\Delta\omega^2 \left(\frac{1}{2} - \frac{1}{2}\cos(2\omega_{sc}t)\right)\right). \quad (4.30)
$$

In the ideal case the offset frequency is zero, $\omega_{off} = 0$, and the only remaining useful term is the term at twice the sub-carrier frequency

$$
s_{dem,2}(t) = \frac{A^2}{16}\Delta\omega^2\tau_0^2 \cos(2\omega_{sc}t). \quad (4.31)
$$

The demodulation can now be performed using this signal. The same conclusion holds for the triangular wave since it can be represented by the Fourier series, in which case again, the second harmonic of the demodulated sub-carrier wave can be used for the final FSK demodulation. Interestingly, if an ideal differentiator is used and if infinite IF bandwidth is assumed, the amplitude of the second harmonic will be independent of the frequency offset. The first harmonic will appear with the increase of the frequency offset, however in this case, this component can be filtered out by the LF band-pass filter.

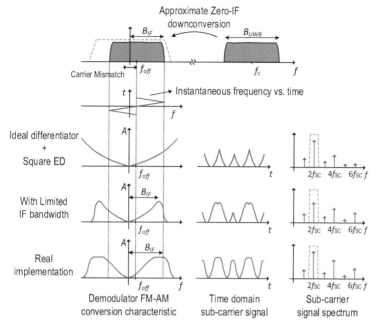

Figure 4.7 Principle of operation of the approximate zero IF receiver with single-ended downconversion.

The principle of the single-ended baseband FM demodulator is shown in Figure 4.7. In the derivation, an ideal differentiator was used and infinite IF bandwidth was assumed. In a realistic implementation the IF bandwidth will affect the useful signal and will cause the second harmonic of the demodulated signal (used for demodulation) to decrease with frequency offset. Also, the ideal differentiator, used for derivation, will be replaced by a lossy (non-zero dc gain) first order high-pass filter. This filter will have a certain cut-off frequency after which the transfer function flattens. The equivalent FM-AM characteristic should finally resemble the characteristic at the bottom of the Figure 4.7.

The first and second harmonic of the demodulated signal are plotted in Figure 4.8 as functions of the offset frequency. The calculation is done for a triangular sub-carrier signal with varying IF bandwidth. As explained previously, the first harmonic is close to zero for small frequency offsets, and increases as the offset increases. The second harmonic (useful part of the signal), decreases with the frequency offset. This decrease is purely a consequence of the limited demodulator bandwidth, since the second harmonic

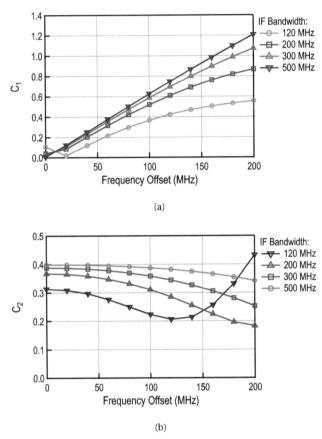

(a)

(b)

Figure 4.8   Normalized fundamental C_1 and second harmonic amplitude C_2 at the output of the demodulator.

after a perfect square law envelope detector remains constant regardless of the offset. The amplitude of the second harmonic shows less variation with the frequency offset as the IF bandwidth increases. Ideally, the IF bandwidth should then be extended to get the best performance, however this again requires more power for the IF amplifiers, and in the receiver implementation a bandwidth of 300 MHz will be used as a good trade-off.

4.5 Receiver Sensitivity Estimation

The first step in estimating the receiver sensitivity is to find the output BER as a function of the SNR at the receiver input. For the approximate zero IF receiver with quadrature demodulation, the FM-AM conversion characteristic is equivalent to the one of the RF delay line demodulator. The only difference between the two is that the demodulator is located at the baseband instead of RF. The expectation is then that the BER performance of the two receivers remains the same, meaning that the same approximation can be used to estimate the BER. The hypothesis is verified using the high-level model corresponding to the one shown in Figure 4.2. A bandwidth of 300 MHz was used for the IF filters, and 2 MHz for the LF filter that filters the demodulated FSK signal. The bandwidth of the demodulator is chosen larger than the IF bandwidth and is set to aproximately 350 MHz. The simulation results are compared to the Gerrits' approximation [8] in Figure 4.9, using the formula for a non-coherent probability of error. The simulation points match well with the calculated curve, validating the use of Gerrits' approximation for the proposed quadrature receiver.

In the case of the single-ended receiver architecture, the Gerrits' approximation no longer holds in it's original form. However, looking at the receiver structure, after the squaring operation of the envelope detector, the same products appear as in the case of the delay line demodulator. In principle the same approach can be used as in [8], with the difference that the useful signal amplitude is half of the one in the case of the delay line demodulator. The resulting output SNR is then given by

$$\text{SNR}_{\text{out}} = \frac{B_{RF}}{B_{SC}} \frac{(\text{SNR}_{\text{in}}/4)^2}{1 + \text{SNR}_{\text{in}}}. \tag{4.32}$$

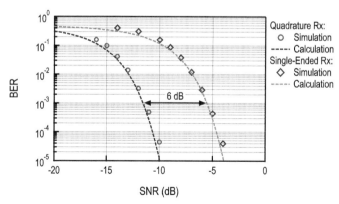

Figure 4.9 Simulated and calculated BER curves for the approximate zero IF receiver.

The rest of the calculation remains the same as for the delay line demodulator. The resulting calculated BER curve is shifted by approximately 6 dB compared to the BER of the quadrature receiver. This is the price paid for the simplified receiver architecture. The calculated BER curve is compared to the simulated points in Figure 4.9. The used model corresponds to the block diagram of Figure 4.6. The IF bandwidth of 300 MHz was used in the simulation, and a high-pass filter with a cut-of frequency of 300 MHz replaced the differentiator. Such implementation should roughly correspond to the actual implementation of the single-ended receiver.

The two BER curves provide the information on the minimum SNR needed at the receiver input in order to achieve the desired BER. Typically, for low power wireless receivers a BER of 10^{-3} is taken as the reference point. The sensitivity of the receiver is then defined as the signal level at the receiver input needed to achieve this BER. To calculate this sensitivity, the noise power at the receiver input must first be calculated. The equivalent input referred noise of the receiver in dBm is given by

$$N = 10\log(\mathrm{k}TB_{UWB}/1\,\mathrm{mW}) + NF, \qquad (4.33)$$

where NF is the noise figure of the receiver, and $10\log(\mathrm{k}TB_{UWB}/1\,\mathrm{mW})$ is the thermal noise power in dBm at the receiver input at a temperature of T $= 25°$C. The used FM-UWB signal bandwidth is $B_{UWB} = 500\,\mathrm{MHz}$. Considering that the main target is to lower the receiver power consumption and that the LNA will be either completely removed, or have very limited performance, relatively high noise figure of the receiver should be accounted for. In [5] the total noise figure of the active mixer and the IF amplifiers is 23 dB. The high noise figure is a consequence of the mixer first architecture and low power gain of the first stage, which results in significant contribution from the IF amplifier. For this design, a 20 dB noise figure will be assumed in order to calculate the achievable sensitivity of the FM-UWB receiver. The sensitivity is then calculated as

$$S_{in} = 10\log(\mathrm{k}TB_{UWB}/1\,\mathrm{mW}) + NF + \mathrm{SNR}_{min}. \qquad (4.34)$$

For the quadrature receiver minimum input $\mathrm{SNR}_{min} = -11.5\,\mathrm{dB}$, which results in a receiver sensitivity of around $S_{in} = -78.5\,\mathrm{dBm}$. For the single-ended receiver the minimum input SNR is approximately 6 dB higher, which results in sensitivity of $S_{in} = -72.5\,\mathrm{dBm}$. Achievable sensitivity, although low compared to typical narrowband receivers that achieve levels lower than $-90\,\mathrm{dBm}$ (for example, typical Bluetooth receivers), is sufficient for communication in body area networks at distances below 1 m. The presented calculation and simulation are valid in an ideal case, where the LO frequency is perfectly aligned with the center frequency of the FM-UWB signal. Since the idea behind power reduction is to use a low quality oscillator whose frequency might drift with time, the sensitivity degradation due to frequency offset should be estimated as well. This is done using the same high-level model, and the results are presented in Figure 4.10.

The 50 MHz offset is taken as a maximum offset that should be tolerated, and the LO frequency must be maintained within these limits. In the practical implementation this will be achieved using a calibration FLL loop

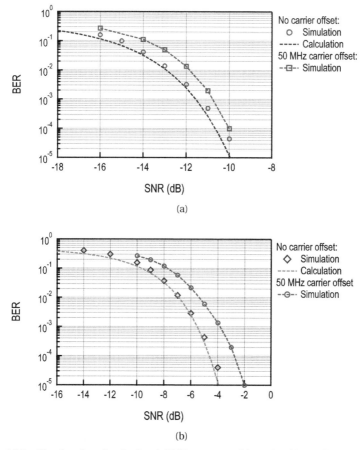

Figure 4.10 Simulated and calculated BER curves with and without frequency offset for the approximate zero-IF receiver with quadrature downconversion (a) and single-ended downconversion (b).

that is periodically turned on (e.g. by a microcontroller when temperature exceeds certain limits). The sensitivity degradation for a 50 MHz frequency offset and for the receiver with quadrature downconversion is reported in Figure 4.10(a). As can be seen, this degradation is below 1 dB, confirming empirically the initial hypothesis that the frequency offset does not cause a major sensitivity degradation. In the case of the receiver with single-ended downconversion this degradation amounts to around 1.6 dB, as seen in Figure 4.10(b) showing that the simplified architecture is slightly more susceptible to frequency offsets.

It should be noted here that the given calculation and simulation models only account for a noise source at the input of the receiver. The separate contributions of receiver blocks are accounted for through the noise figure, however what is not accounted for is the noise generated by the FM demodulator (delay line demodulator for the quadrature, and envelope detector for the single-ended receiver). Due to the nonlinear nature of FM demodulators, the output noise will depend on the input signal level. In order for the presented sensitivity estimation to be valid, the noise of this block must be negligible compared to the noise from other sources. This requirement is simply achieved by increasing the gain of the stages preceding the demodulator, and it is in fact this requirement that sets a limit for the combined gain of the LNA, mixer and the IF amplifier.

4.6 Summary

This chapter presents the general approach to receiver power reduction through use of "uncertain IF" and "approximate zero IF" architectures. The main idea is to loosen constraints on RF stages, that usually consume the most power, and shift the burden to IF where high gain comes at a lower price in terms of power. Two different receiver architectures are proposed. The quadrature approximate zero IF receiver targets to reduce consumption, but also to provide enough linearity to support multi-user communication. Potential to parallelize communication through sub-carrier FDMA, on top of existing TDMA could bring both latency and power savings at a network level. The second architecture, the single-ended FM-UWB receiver architecture, aims solely to reduce power consumption. The used approach sacrifices all other performance aspects in order to provide the lowest possible consumption level, and could be used when there is no need for SC-FDMA. The analysis of the two architectures is presented, providing some insight into the key points and the principle of operation, together with a short sensitivity analysis that estimates the achievable receiver performance.

The implementation of the concepts presented here is the subject of the following chapters. First, the quadrature approximate zero IF FM-UWB receiver is implemented and characterized standalone. Then, in the second iteration, a full transceiver is integrated. Both receivers are placed on the same die, with the idea to use the single-ended FM-UWB receiver as a low power mode.

References

[1] A. Sai, H. Okuni, T. T. Ta, S. Kondo, T. Tokairin, M. Furuta, and T. Itakura, "A 5.5 mW ADPLL-based receiver with a hybrid loop interference rejection for BLE application in 65 nm CMOS," *IEEE Journal of Solid-State Circuits*, vol. 51, no. 12, pp. 3125–3136, Dec. 2016.

[2] N. Saputra and J. R. Long, "A short-range low data-rate regenerative FM-UWB receiver," *IEEE Transactions on Microwave Theory and Techniques*, vol. 59, no. 4, pp. 1131–1140, Apr. 2011.

[3] F. Chen, W. Zhang, W. Rhee, J. Kim, D. Kim, and Z. Wang, "A 3.8-mW 3.5-4-GHz regenerative FM-UWB receiver with enhanced linearity by utilizing a wideband LNA and dual bandpass filters," *IEEE Transactions on Microwave Theory and Techniques*, vol. 61, no. 9, pp. 3350–3359, Sep. 2013.

[4] Y. Zhao, Y. Dong, J. F. M. Gerrits, G. van Veenendaal, J. Long, and J. Farserotu, "A short range, low data rate, 7.2 GHz-7.7 GHz FM-UWB receiver front-end," *IEEE Journal of Solid-State Circuits*, vol. 44, no. 7, pp. 1872–1882, July 2009.

[5] N. M. Pletcher, S. Gambini, and J. Rabaey, "A 52 μW wake-up receiver with -72 dBm sensitivity using an uncertain-IF architecture," *IEEE Journal of Solid-State Circuits*, vol. 44, no. 1, pp. 269–280, Jan. 2009.

[6] J. Gerrits, J. Farserotu, and J. Long, "A wideband FM demodulator for a low-complexity FM-UWB receiver," in *The 9th European Conference on Wireless Technology, 2006*, Sep. 2006, pp. 99–102.

[7] M. Kouwenhoven, *High-Performance Frequency-Demodulation Systems*. Delft University Press, 1998.

[8] J. F. M. Gerrits, M. H. L. Kouwenhoven, P. R. van der Meer, J. R. Farserotu, and J. R. Long, "Principles and limitations of ultra-wideband FM communications systems," *EURASIP J. Appl. Signal Process.*, vol. 2005, pp. 382–396, Jan. 2005.

[9] V. Kopta, D. Barras, and C. C. Enz, "An approximate zero IF FM-UWB receiver for high density wireless sensor networks," *IEEE Transactions on Microwave Theory and Techniques*, vol. 65, no. 2, pp. 374–385, Feb. 2017.

[10] J. Farserotu, J. Baborowski, J. D. Decotignie, P. Dallemagne, C. Enz, F. Sebelius, B. Rosen, C. Antfolk, G. Lundborg, A. Björkman, T. Knieling, and P. Gulde, "Smart skin for tactile prosthetics," in *2012 6th International Symposium on Medical Information and Communication Technology (ISMICT)*, Mar. 2012, pp. 1–8.

5

Quadrature Approximate Zero-IF FM-UWB Receiver

5.1 Introduction

The previous chapters explained the basics of FM-UWB modulation, discussed the existing state of the art and introduced two new architectures for an FM-UWB receiver. The concept of the proposed approximate zero-IF architecture with quadrature downconversion is brought to life in this chapter. The work is mainly oriented toward exploiting the short communication range, in order to lower power consumption of the receiver, but also to provide means to efficiently communicate as the number of sensor nodes in the network scales up. This is achieved through the use of the sub-carrier FDMA, that allows to distinguish multiple FM-UWB signals sharing the same RF band.

The chapter starts by introducing the top-level architecture of the integrated receiver. The following section deals with the details of circuit design, focusing on the key approaches and techniques used to reduce the power consumption of the most important circuits. Measurements of the implemented receiver are presented in Section 4.3. Beyond the intended data rate of 100 kb/s, the receiver is characterized in additional scenarios (higher speed, M-FSK modulation, multi-channel transmission) showing the true potential of the FM-UWB modulation. Finally, the chapter is concluded with a summary of achieved results and a comparison with similar receivers from the literature.

5.2 Receiver Architecture

The aim of this work is to reduce power of the FM-UWB receivers beyond the current state of the art while preserving the demodulator linearity and multi-user communication capability. As the LNA and other blocks operating at RF have been shown to consume the most power in previously implemented

89

receivers, the strategy here is to first downconvert the signal to baseband, and then perform all the processing at low frequencies. Since power consumption in all of these blocks typically increases with frequency, moving them to baseband should result in significant power savings.

An oscillator, that was not needed in previous FM-UWB receiver implementations, is now necessary to generate the LO signal. Only if the LO can be implemented with a reasonable power budget can the approximate zero-IF architecture lower the overall consumption. Fortunately, ring oscillators in deep sub-micron technology nodes are known to consume very little and can be used here for such LO generation. The downside of using a low power ring oscillator is its high phase-noise, unstable oscillation frequency and high susceptibility to environmental changes. Ring oscillators are almost exclusively used in closed loop systems such as PLLs, where the oscillator is locked to a reference frequency, and all the aforementioned problems disappear. However, a PLL would require frequency dividers and these would add a significant contribution to the receiver power consumption. Instead of implementing a PLL, a free-running oscillator is used for this implementation and owing to the large FM-UWB signal bandwidth, some of the issues, such as phase noise, are circumvented. Frequency dividers are still needed, but they are used as a part of the FLL calibration loop and are turned on only when calibration is necessary. That is mostly to compensate for frequency drift due to temperature or supply voltage variations. Fortunately, since these changes are slow and the calibration is not needed very often, the FLL calibration circuits will not pose a significant overhead to the receiver consumption.

The high-level block diagram of the implemented receiver is shown in Figure 5.1, with the main receiver at the bottom and the test receiver at the top. The two receivers are implemented in order to asses the performance loss due to the on-chip, low power ring oscillator. They are identical in all aspects except for the LO. The main receiver uses the ring oscillator, whereas the test receiver uses an external signal to drive the mixer. Since this is a direct conversion receiver, quadrature LO signals are generally needed to perform correct demodulation. In the main receiver these are generated directly by the ring (multiple stages produce different phases, as will be shown later), while the test receiver uses an RC-CR network to provide quadrature signals, allowing to reduce the number of input pads. The difference in performance between the two receivers will be reported in the measurement section.

As already mentioned, the idea is to reduce power consumption by removing the RF blocks. It can be seen in Figure 5.1 that the LNA is still present, however, in this implementation it is simply a transconductance amplifier

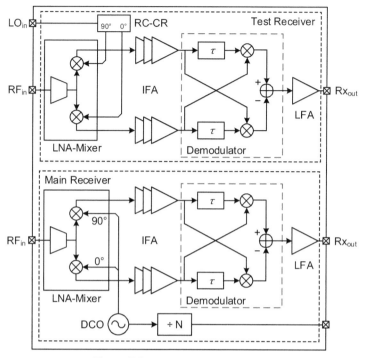

Figure 5.1 Receiver block diagram.

that converts the input voltage into current that is then downconverted by the mixer. The two can also be regarded as an active mixer with an input matching network. Since this is practically a mixer-first receiver the noise figure will be higher compared to a more standard approach with the LNA in front, but this remains an acceptable price for the achieved power savings. The main gain stages are placed at baseband (here referred to as IF amplifiers), allowing to achieve higher gain at lower power. Since the oscillation frequency of the ring oscillator is not stable, the bandwidth of these amplifiers is increased to account for a ± 50 MHz carrier frequency offset. Instead of 250 MHz that would normally be sufficient to amplify a downconverted 500 MHz wide signal, the bandwidth of IF amplifiers is extended to 300 MHz. The IQ delay line demodulator is a modified version of the demodulator from [1], adjusted for baseband operation, as described in the previous chapter.

The receiver presented here only implements the first FM demodulation. The resulting demodulated sub-carrier signal is buffered and is available at the receiver output. This signal is then converted to digital domain using an ADC, allowing the further data processing to be conducted off-line. The second FSK demodulation, and all the additional baseband processing (e.g. channel filtering) is implemented in software, allowing to measure BER performance of the receiver. It should be noted that this idealized approach yields a somewhat better performance than otherwise achievable with a low-power hardware implementation, but can nevertheless be used to assess the performance of the integrated blocks. All of the implemented circuits can be controlled through an SPI bus, allowing to tune the bias current, resonance frequency, gain and bandwidth of different blocks and switch them on or off. Details of circuit implementation are given in the following section.

5.3 Circuit Implementation

5.3.1 RF Frontend

The LNA and the mixer, shown in Figure 5.2, are stacked in order to save power. The circuit can also be seen as an active mixer with the input matching network. An active mixer is chosen for downconversion because unlike a passive mixer, it provides voltage gain and does not require a rail to rail LO swing, preventing excessive consumption in the LO buffers. The used LO swing is around 300 mV peak to peak (single-ended), which is sufficient for the chosen circuit topology. Increasing the swing to 1 V, would result in an increase of the LO buffer power consumption by more than a factor of 9 (proportional to V_{LO}^2), hence justifying the choice of an active mixer. The transistor M_1 acts as a main transconductance stage, that converts the input voltage into current before the downconversion. Center-tapped symmetric inductor L_1 acts as a transformer and boosts the equivalent transconductance of transistor M_1 [2, 3], without the increase of power consumption, making this approach ideal for a low power design. Disregarding capacitor C_T for the moment, and assuming C_2 is large enough to be considered as a short circuit at the frequencies of interest, the equivalent transconductance seen from the gate of M_1 is given by

$$G_{m,eq} = \frac{\Delta I_1}{\Delta V_G} = \frac{(k+1)G_{m1}}{1 + j\omega L G_{m1}(1 - k^2)}, \qquad (5.1)$$

where k is the transformer coupling coefficient. As k approaches 1 (ideal transformer) the equivalent transconductance becomes purely real and equal to $2G_{m1}$. It has been shown that, for the same current consumption, this

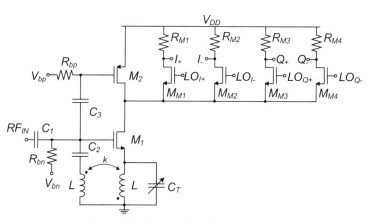

Figure 5.2 Schematic of the LNA/Mixer.

approach achieves higher gain and larger bandwidth compared to the standard inductively degenerated LNA [3].

It was shown in [2] that the input impedance of the transformer feedback LNA is given by (assuming C_2 is large, $C_{GS} \ll C_1$, $k = 1$, $C_T = 0$ and $\omega^2 L^2 G_{m1}^2 \gg 1$)

$$Z_{in} \approx \frac{1}{j\omega C_1} + j\omega L + 2\omega^2 L^2 G_{m1}. \tag{5.2}$$

Resonance frequency is then $\omega_0^2 = 1/LC_1$. In this design, tuning capacitor C_T is added to provide capability to tune the resonance frequency and to compensate for small component variations. With C_T the expression of the input impedance becomes:

$$Z_{in} \approx \frac{1}{j\omega C_1} + j\omega L \frac{1 - \omega^2 LC_T(2 - k^2)}{1 - 2\omega^2 LC_T}$$
$$+ \omega^2 L^2 G_{m1} k(1 + k) \frac{1 - \omega^2 LC_T(1 - k)}{1 - 2\omega^2 LC_T}. \tag{5.3}$$

In the above expression it is assumed that $C_T \ll C_1$, which means that close to resonance $\omega^2 LC_T < 1$. Assuming $k = 1$, 5.3 can be further simplified to:

$$Z_{in} \approx \frac{1}{j\omega C_1} + j\omega L \frac{1 - \omega^2 LC_T}{1 - 2\omega^2 LC_T} + \frac{2\omega^2 L^2 G_{m1}}{1 - 2\omega^2 LC_T}, \tag{5.4}$$

which shows that the resonance frequency is now a function of C_T. Unfortunately, C_T also affects the real part of the input impedance, however it is still possible to achieve good matching and roughly 10% tuning range of the resonance frequency.

A common problem in active mixers is that the bias current required by the transconductance M_1 and switching transistors M_{M1-4} is not the same. Bias current of M_1 is set by the input matching condition and the desired voltage gain. Voltage gain of the active mixer is proportional to the product of the transconductance and the load resistance $G_{m1}R_{M1-4}$. At the same time, the dc point of the output voltage and the LO feedthrough ($I_{+/-}$ and $Q_{+/-}$ outputs) are dependent on the product of the bias current and load resistance $I_b R_{M1-4}$. Increasing the voltage gain, either through G_{m1} (and consequently I_b) or through R_{M1-4} lowers the output bias voltage and increases the LO feedthrough. To add a degree of freedom and break this dependence, "current stealing" technique can be used. This is accomplished using the transistor M_2, that sinks part of the M_1 bias current. In this way, mixer bias current can be set independently of the M_1 bias current, allowing to break the dependence between the voltage gain on one side and dc bias and LO feedthrough on the other. As a consequence, load resistor values can be increased to maximize voltage gain without causing excessive LO feedthrough. In addition to current stealing, since the gate of M_2 is connected to the LNA input through a large capacitor C_3, it also contributes to the overall transconductance, further increasing voltage gain. The approach is similar to the complementary LNA presented in [4], with the difference that the bias currents of M_1 and M_2 are not the same. The addition of M_2 has some downsides in a practical implementation. More complex layout of the LNA will result in increased parasitics, and more importantly drain capacitance of M_2 will be added to the parasitic capacitance at the mixer input, effectively reducing bandwidth of the RF front-end. To compensate for the added capacitance, the equivalent input resistance of the mixer can be reduced by increasing the size of the switching transistors M_{M1-4}, but this comes at price of increasing the load of the LO buffers.

In this design, resistors R_{M1-4} can be switched between $22\,\mathrm{k\Omega}$ and $14\,\mathrm{k\Omega}$ and provide two gain steps for the mixer. Since the voltage gain is obtained entirely at baseband frequencies, after mixing, it comes at a lower cost in terms of power, and eliminates the need for a resonant load, thereby saving silicon area. Achieved voltage gain is around $15\,\mathrm{dB}$ over a $600\,\mathrm{MHz}$ bandwidth, for the maximum gain setting. Simulated current consumption is $70\,\mu\mathrm{A}$ from a $1\,\mathrm{V}$ supply. The input referred $1\,\mathrm{dB}$ compression point

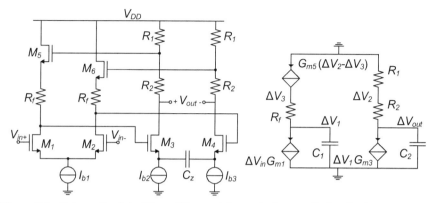

Figure 5.3 Schematic of the IF amplifier, and the equivalent small-signal schematic of half circuit.

of the RF frontend is at $-19\,\mathrm{dBm}$. The price to be paid for low power consumption is elevated noise figure, which in this case is $15\,\mathrm{dB}$ according to simulations (this is including the LNA and the mixer). Even though it is higher than the typical values found in standard receiver implementations, it is still acceptable for communication over short distances.

5.3.2 IF Amplifier

The mixer is followed by the I and Q IF amplifiers that provide most of the voltage gain. Each IF amplifier is a cascade of three modified CMOS Cherry-Hooper (CH) amplifiers shown in Figure 5.3. The basic concepts that come from [5] were further developed in [6], where emitter-follower was introduced in the feedback, and the first CMOS version was presented in [7]. A CH amplifier is a feedback amplifier with a second order transfer function. Compared to a cascade of standard differential pairs, feedback amplifiers offer larger bandwidth for the same power consumption. This is why these amplifiers were originally used for high data rate optical receivers, targeting bandwidths of more than $1\,\mathrm{GHz}$. In this case, the design was optimized for $300\,\mathrm{MHz}$ bandwidth and low power consumption. By controlling the Q-factor of the transfer function, behavior close to the edge of the pass-band can be controlled. In this particular case peaking was used to compensate for the slight drop in the LNA/mixer conversion gain close to the band edges and provide a relatively flat overall gain characteristic.

The small-signal schematic of the half-circuit is given in Figure 5.3. Capacitors C_1 and C_2 are a combination of gate capacitance (in the case of C_2 this would be the gate capacitance of the following stage) and layout parasitics. Capacitance C_z introduces a zero in the transfer characteristic, and is used to prevent offset accumulation in the IF amplifiers. Although, strictly speaking, the downconverted FM-UWB signal occupies frequencies from 0 to 250 MHz, a zero in the transfer function will not affect the performance of the demodulator as long as this zero is low compared to the signal bandwidth. In this case the zero is placed around 1 MHz, and since it will not affect the behavior in the pass-band it is not considered in the small-signal analysis. Gain in the pass-band is given by [7]:

$$A_{v0} = \frac{G_{m1}(R_1 + R_2)(1/G_{m5} + R_f)}{(1/G_{m3} + R_1)}. \tag{5.5}$$

Assuming $G_{m5}R_f \gg 1$ and $G_{m3}R_1 \gg 1$ the above expression reduces to

$$A_{v0} \approx \frac{G_{m1}(R_1 + R_2)R_f}{R_1}. \tag{5.6}$$

As the voltage gain is a function of the ratio of the two load resistors R_1 and R_2, gain switching can be implemented by switching the value of R_2. The second order transfer function of the CH amplifier is given by

$$A_v(s) = \frac{G_{m1}G_{m3}(R_1 + R_2)(1 + G_{m5}R_f)}{a + bs + cs^2}, \tag{5.7}$$

$$a = G_{m5}(1 + G_{m3}R_1),$$
$$b = (C_1(1 + G_{m5}R_f) + G_{m5}C_2(R_1 + R_2)),$$
$$c = C_1C_2(R_1 + R_2)(1 + G_{m5}R_f).$$

Again, assuming the transconductances are high enough that $G_{m5}R_f \gg 1$ and $G_{m3}R_1 \gg 1$ leads to the simplification of the expression that reduces to

$$A_v(s) = \frac{G_{m1}G_{m3}(R_1 + R_2)R_f}{G_{m3}R_1 + s(C_1R_f + C_2(R_1 + R_2)) + s^2C_1C_2(R_1 + R_2)R_f}. \tag{5.8}$$

Capacitors C_1 and C_2 are determined by the size of the differential pair transistors and the layout parasitics. Gain, bandwidth and Q-factor of the amplifier transfer function are then set by the resistances of R_1, R_2 and R_f, which can be used as design parameters.

Simulated gain of the standalone LNA/mixer, and the LNA/mixer together with IF amplifiers is shown in Figure 5.4(a). Overall gain of all the

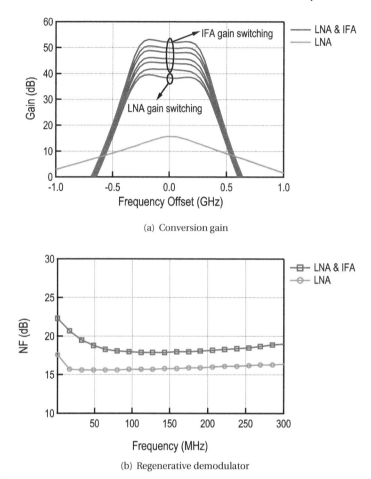

(a) Conversion gain

(b) Regenerative demodulator

Figure 5.4 Simulated conversion gain and noise figure of the RF and IF stages.

stages preceding the FM demodulator is around 53 dB, with approximately 38 dB provided by the IF amplifier. Each CH amplifier cell requires 20 μA of current, which results in 120 μA consumed by the I and Q IF amplifier chains. Equivalent 6th order filtering characteristic provides the attenuation of 32 dB at an offset frequency of 500 MHz. Gain control is implemented through switching of R_2, that can take one of the values 6 kΩ, 18 kΩ and 30 kΩ, while $R_1 = 24$ kΩ. With three cascaded stages, the designed IF amplifier provides 6 different gain levels and one additional level is provided by switching the mixer load resistors R_{M1-4}. Different gain levels can be seen in Figure 5.4(b). Figure 5.4(b) shows the simulated noise figure of the standalone RF frontend and of the RF frontend and the IF amplifiers together. The RF frontend provides around 15 dB of voltage gain, however the power remains low. This is a consequence of low bias current of the LNA/mixer, that results in a low value of transconductance. As a result, the noise added by the IF amplifiers will increase the total noise figure by approximately 3 dB. The noise figure of the standalone IF amplifier is around 5 dB in the pass-band. Finally, even though the noise figure is higher compared to more conventional receiver implementations, the achieved levels still provide enough sensitivity for communication over short distances.

5.3.3 LO Generation and Calibration

The proposed receiver is intended for use in the lower part of the UWB band, targeting 500 MHz wide signal centered around 4 GHz. The emphasis of the work described here is on reducing the power consumption of the receiver, while still preserving the capability to operate in an environment where several FM-UWB transceivers might be communicating at the same time. The power reduction dominantly comes from the fact that the gain stages operate at low frequencies, while no voltage gain is provided at RF. However, such an approach can only be beneficial if the LO signal can be generated efficiently. Additional difficulty is the need for quadrature signals since a 90° shift is generally required for correct demodulation in a direct conversion (zero IF) receiver. Providing such signals at 4 GHz tends to be power costly. Fortunately, owing to the properties of FM-UWB and the chosen receiver architecture, the oscillator constraints are quite loose. Due to the large bandwidth of the FM-UWB signal, phase noise is not a major concern (-80 dBc at 10 MHz offset according to [8]) and no precise frequency generation is needed, therefore a simple free-running ring oscillator can be used to provide carrier signals for downconversion. When it comes to power consumption, ring oscillators are advantageous compared to LC oscillators, as they benefit from technology scaling. Inductor quality factor, which is a limit to power consumption of integrated LC oscillators, remains constant and practically independent of technology. On the other hand, gate capacitance and interconnect parasitic capacitances, that determine consumption of ring oscillators, decrease with technology scaling. This enables the reduction of power consumption of the ring oscillator, making the proposed approach favorable for future implementations.

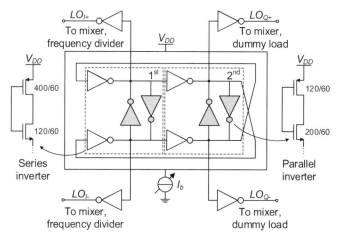

Figure 5.5 Simplified schematic of the quadrature DCO.

The oscillator schematic is shown in Figure 5.5. A chain with an even number of inverters is needed to provide quadrature signal generation, however such a circuit would latch on its own. In order to prevent latching, and force a 180° phase shift, parallel inverters are added between the corresponding nodes (grey inverters in Figure 5.5). A different way to see the implemented oscillator would be as a pseudo-differential two stage ring oscillator, where each stage consists of four inverters and the differential mode is enforced by the parallel inverters [9, 10]. The two stages provide a 90° phase shift, and an additional 180° shift is provided by cross-coupling the stages, hence assuring a reliable start-up. The series and parallel inverters are sized differently, W/L ratios in nm are shown in Figure 5.5, dimensions were optimized for low power consumption. One of the difficulties in designing very low power ring oscillators is that capacitive load is dominantly determined by the capacitance of the interconnect wires, and is layout dependent. Careful layout design with several iterations is needed to minimize power consumption. Correct phase relations between different signals are guaranteed by symmetry, however a small quadrature error is present due to mismatch between transistors. The error will vary from die to die and according to Monte Carlo simulations their standard deviation is $\sigma_\phi = 2.6°$. Frequency is controlled via supply current of the current starved CMOS inverters. All inverters share the same current source as this approach was proven to perform better than the solution with a separate current source for each inverter, or inverter pair [9].

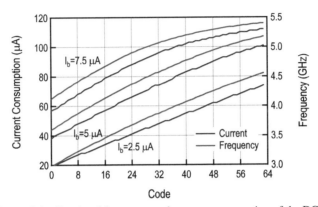

Figure 5.6   Simulated frequency and current consumption of the DCO.

Since process-voltage-temperature (PVT) variations can have a significant impact on the oscillation frequency, the digitally controlled oscillator (DCO) was designed to cover the frequency range from 3 GHz to 5 GHz, thus assuring that it can be tuned correctly under all conditions. A 6-bit current DAC is used to provide the supply current, resulting in less than 30 MHz frequency resolution. The frequency step is not constant due to non-linear characteristic of the DCO and decreases as the oscillation frequency increases. At 4 GHz the DCO produces a 300 mV peak-to-peak single-ended signal while consuming 140 μA (including the buffers). Simulated oscillation frequency and current consumption of the DCO, as functions of input code, are shown in Figure 5.6 for different DAC reference currents.

Since the oscillation frequency of the DCO is imprecise and prone to environment changes it must be calibrated periodically to assure correct operation (e.g. to compensate for temperature). Since environmental changes are slow, the calibration would only need to be done once in a few hours or potentially even days, meaning that the consumption of the calibration circuitry on average remains negligible compared to the receiver consumption. The calibration can be done using a frequency-locked loop (FLL) that is turned on as needed. The FLL was not integrated in this implementation, however it can be added externally using a microcontroller or an FPGA, and the available output from the on-chip frequency divider. A fixed ratio, integer frequency divider is implemented as a cascade of 10 divide-by-2 cells. By selecting outputs from different dividers, one of the four divide ratios 128, 256, 512 and 1024 can be selected as an output for calibration. Each cell is a simplified version of a dynamic 2/3 divider circuit described in [11]. It was designed to cover a somewhat larger range of frequencies than the DCO

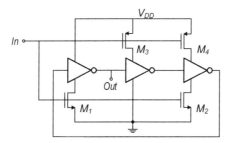

Figure 5.7   Schematic of the frequency divider.

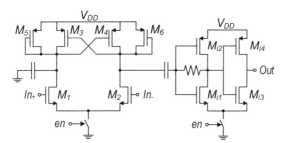

Figure 5.8   Schematic of the buffer between the DCO and the frequency divider.

to assure reliable operation. Owing to the simplified structure, the circuit from Figure 5.7 can work up to 6 GHz. Since the divider requires a rail-to-rail input signal, it is preceded by a buffer from Figure 5.8 that performs differential to single-ended conversion and amplifies the signal. The first stage of the buffer is a pseudo-differential amplifier that uses positive feedback to boost the gain. The positive feedback is implemented using the cross-coupled transistors M_3 and M_4 that provide a negative transconductance. This negative transconductance is used to minimize the equivalent output conductance of the amplifier and increase gain. The differential amplifier is followed by inverters that further amplify the LO signal and produce a rail-to-rail voltage at the output. The whole buffer consumes around 250 μW at 4 GHz and its power consumption is proportional to the input frequency. Figure 5.9 shows simulated waveforms at the buffer input and output, and divider signals in different points. The whole divider chain consumes around 150 μW, and the largest part of the consumption is coming from the first two stages that operate at the highest frequencies.

The frequency divider buffer itself is connected to the LO_{I+} and LO_{I-} outputs of the DCO buffers. Dummy load is added to LO_{Q+} and LO_{Q-} to prevent amplitude mismatch between the I and Q LO signals. Even though

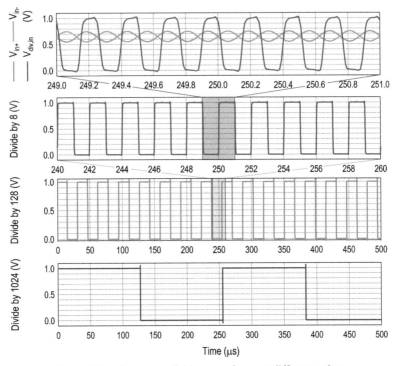

Figure 5.9   Frequency divider, waveforms at different points.

the DCO itself can produce a sufficiently large output amplitude, the four buffers (Figure 5.5) need to be placed between the internal nodes of the DCO and the inputs of the mixer and divider. This is done to decouple the oscillation frequency from the mode of operation (reception or calibration). The input capacitance of the divider buffer varies with the bias current of the two input transistors, and is different in the on and off state. If these buffers were connected directly to the DCO, the change in load capacitance would cause a shift in frequency after calibration. In addition, the presence of DCO buffers reduces coupling between the external signal and the DCO, thus preventing the pulling effect (shift in frequency caused by external signal). The four buffer inverters consume around $80\,\mu A$, almost 60% of the entire DCO consumption which is a significant but necessary overhead.

5.3.4 FM Demodulator

The implemented wideband FM demodulator is a modified version of the delay-line demodulator presented in [1]. In order to conserve power, the demodulator has been moved from RF to baseband, but it now requires quadrature inputs to perform correct FM demodulation. Additional benefit of moving the demodulator to baseband is that there is no need for precise delay generation, as it is no longer related to the input signal frequency. The only limit is coming from the demodulator bandwidth that is inversely proportional to the delay. The implemented delay-line demodulator is presented in Figure 5.10. Two double-balanced Gilbert's mixers perform multiplication of the I and Q signals with their delayed copies. The output currents of the two cells are combined to implement subtraction and produce the demodulated signal. Top inputs of Gilbert's mixer are connected directly to the outputs of the IF amplifier (gates of transistors M_{3-6}). Source followers are placed between the IF amplifier and bottom inputs (gates of transistors $M_{1,2}$) to provide a correct dc level of the input voltage. The delay path consists of source followers $M_{SF1,2}$ and bottom transistors of the Gilbert's mixer $M_{1,2}$. For a first order filter with a pole at ω_p it can be shown that the delay through the filter is equal to $1/\omega_p$, for a signal whose maximum frequency is sufficiently below the cut-off frequency. In this case the total delay is a sum of delays that come from two poles. The first one is associated to the source

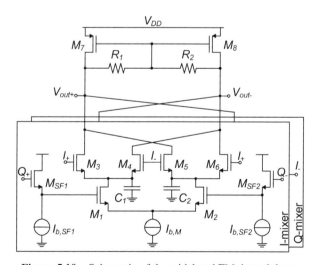

Figure 5.10   Schematic of the wideband FM demodulator.

follower and is given by

$$\tau_1 \approx \frac{C_{G1}}{G_{m,SF1}}, \tag{5.9}$$

where the capacitance C_{G1} accounts for the total capacitance seen at the gate of M_1. Since M_1 is a relatively small transistor, parasitics will contribute a significant portion of the total load.

Due to the asymmetry of the double balanced mixer with respect to the two signal paths, some delay will inherently exist between the bottom and top inputs. This delay is caused by the pole that exists due to the parasitic capacitance at the drain of M_1. To provide better control of the delay and reduce dependence on parasitics, an additional MOM (metal-oxide-metal) capacitor is added in this node ($C_{1,2}$), resulting in the delay that is given by

$$\tau_2 \approx \frac{C_1}{G_{m3} + G_{m4}}. \tag{5.10}$$

All the transistors in the FM demodulator are biased in weak inversion. Since the transconductance of each transistor is proportional to the bias current, delay of the demodulator can consequently be controlled by the bias currents $I_{b,SF}$ and $I_{b,M}$. Two bit control of the bias current is provided to allow delay tuning after production. Load resistors R_1 and R_2 can be switched between the two values to provide two gain settings. The FM demodulator input and output waveforms are shown in Figure 5.11. The figure shows the input sub-carrier signal (top), the I and Q signals (middle) and the demodulated signal (bottom). A small distortion can be seen at the peak values of the demodulated signal. This is a result of the IF amplifier bandwidth, combined with the fact that delay decreases at higher frequencies.

The demodulator consumes only 25 μW, mainly due to the fact that it operates at baseband. Compared with the demodulators from [12, 13], that require close to 6 mW, this is an improvement by two orders of magnitude, allowing significant power savings and still providing sufficient linearity to handle multiple input FM-UWB signals. Additionally, there is no need for inductors and no need for a complex passive network that provides a precise delay, thus resulting in area savings as well.

5.3.5 LF Amplifier and Output Buffer

For the targeted sensitivity levels, the signal amplitude at the output of the demodulator will be too low. Before it can be digitized and analyzed, the

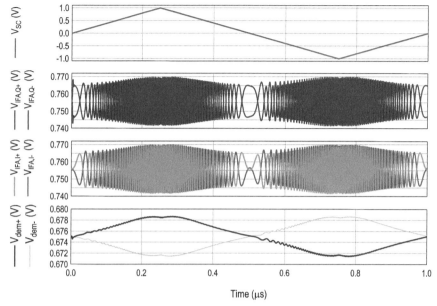

Figure 5.11 Wideband FM demodulator, input and output waveforms.

signal must be amplified and filtered. The two low-frequency (LF) amplifiers that follow the FM demodulator provide a band-pass characteristic from 1 MHz to 2.5 MHz and a maximum voltage gain of roughly 20 dB while consuming 15 μA. As shown in Figure 5.12, each stage is implemented as a fully differential amplifier with resistive source degeneration. Source degeneration provides better linearity, and more precise gain control. The gain of the amplifier is given by

$$A_v = -\frac{G_{m1}R_L}{1 + G_{m1}R_S/2} \approx -\frac{R_L}{R_S/2}. \tag{5.11}$$

The approximation is valid if $G_{m1}R_S/2 \gg 1$, in which case the gain is solely determined by the ratio of load and source resistors. Two bit gain control is provided, both R_L and R_S can be switched between the two values. Since the higher cut-off frequency is determined by R_L and the capacitance loading the amplifier output (gate capacitance of the following stage in series with C_{ac}), decreasing the gain also extends bandwidth. The lower cut-off frequency is determined by the elements of the ac coupling network as $1/R_bC_{ac}$.

Source followers M_{o1} and M_{o2} are placed at the output to provide a low impedance stage that drives the external circuits. They are design to drive

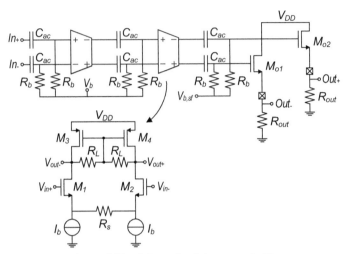

Figure 5.12 Schematic of the output buffer.

a load of 10 pF, which corresponds to the capacitance of the oscilloscope probes or an external ADC, used to digitize the signal, although if necessary an additional external buffer can be added. External resistors define the bias current of the source followers, and can be chosen to have any value between 1 kΩ and 10 kΩ, while still providing sufficient bandwidth.

5.3.6 Current Reference PTAT Circuit

All the circuits described so far require a reference current that defines the bias point. All the reference currents are derived from a single current generated by the circuit from Figure 5.13. The circuit provides a PTAT (proportional to absolute temperature) reference current and reuses the approach from [14]. It is a closed loop circuit made up of two current mirrors, a 1:1 current mirror M_5–M_6 and a 1:K current mirror M_1–M_2. Transistors M_3 and M_4 are used as cascode transistors that define the drain voltage of M_1 and M_2. The bias current is defined by the ratio of M_1 and M_2, and since they are both biased in weak inversion the generated reference current I_{out} is given by

$$I_{out} = \frac{U_T \ln K}{R}. \tag{5.12}$$

The output current is proportional to absolute temperature through thermal voltage $U_T = kT/q$. The generated current is used as a reference current

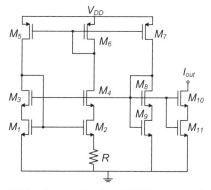

Figure 5.13   Schematic of the PTAT current reference.

for an array of current DACs that provide a reference current for each block of the system. Each of the reference currents can be digitally controlled with a resolution of 1.25 μA, allowing some room for adjustment of the bias current after production.

5.4 Measurement Results

5.4.1 General Receiver Measurements

The proposed receiver was integrated in a standard 65 nm bulk CMOS process. The die photograph is presented in Figure 5.14. The active area of the receiver is approximately 0.4 mm^2, including roughly 450 pF of decoupling capacitance. The receiver only requires one inductor, with no additional off-chip components, which results in smaller area than most existing

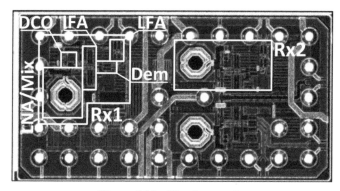

Figure 5.14   Die photograph.

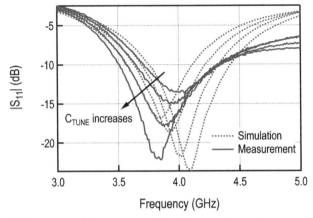

Figure 5.15   Measured S_{11} parameter for different values of input capacitance.

implementations. As already mentioned, two receivers were implemented on the same die for testing purposes. Rx1 is the main receiver that uses the ring oscillator described in the previous section to generate the quadrature LO signals. Rx2 is the test receiver that is identical to the first receiver except that it uses an externally generated LO signal. Two input pads are used for the differential LO, and the on-chip RC-CR network produces quadrature signals. The test receiver is integrated to serve as a reference that allows assessment of performance degradation due to a non-ideal locally generated carrier signal.

Figure 5.15 shows the simulated and measured S_{11} parameter of the receiver for different values of tuning capacitor C_T. A small difference in measured and simulated values is observed. The measurement was done on an FR4 test board with a 10 mm long 50 Ω coplanar waveguide between the pad and the connector. This line was not taken into account in the simulations and might be the cause of the shift in the resonance frequency. Nevertheless, the reflection coefficient is below -10 dB in the band of interest, providing sufficiently good matching.

The DCO frequency was measured using the on-chip frequency divider. As shown in Figure 5.16 the oscillation frequency can be varied from 3.1 GHz to 4.7 GHz. At the same time the supply current of the DCO changes from 32 μA to 85 μA. At 4 GHz the DCO consumes around 60 μA, while the buffer consumes an additional 80 μA, a consequence of the fact that differential quadrature signals need to be buffered. Only one die measurement is presented here, however the frequency characteristic will vary significantly from one die to another as a result of process variation. Nevertheless, all of the measured dies covered the range from 3.6 GHz to 4.4 GHz and could be

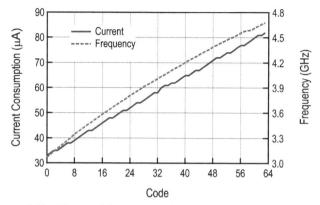

Figure 5.16   Measured frequency and current consumption of the DCO.

Table 5.1   Power consumption breakdown

Block	Current Cons. (μA)	Relative Cons. (%)
LNA & mixer	91	21.5
DCO & buffers	140	33.1
IF amplifier	122	28.8
Demodulator	26	6.1
LF amplifier	15	3.5
Bias	29	6.9

calibrated properly. In all cases the power consumption remains practically the same for the DCO oscillating at 4 GHz. A slightly non-linear behavior can be observed in the output frequency, which is of no significance in this case since the only requirement for the calibration loop is monotonicity, which is satisfied.

The proposed receiver consumes 423 μW of power from a 1 V supply. Power breakdown is shown in Table 5.1. The highest consumer is the DCO together with buffers, followed by the IF amplifiers and the LNA. The demodulator, with only 26 μW of power consumption consumes two orders of magnitude less power than the same type of demodulators implemented previously in [12, 13].

5.4.2 Single User Measurements

The test setup used for the bit error rate (BER) measurements is presented in Figure 5.17. A random bit sequence is generated by software and mapped to the corresponding quadrature FM-UWB symbols. A 12 GS/s, 12-bit arbitrary waveform generator, M8190A was used to generate the baseband quadrature

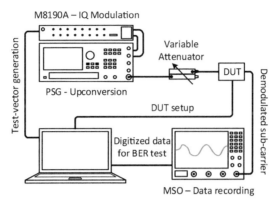

Figure 5.17   Measurement setup.

FM-UWB signals. These signals are subsequently up-converted to 4 GHz by the Keysight PSG signal generator and used for receiver characterization. The flexibility provided by the M8190A allows to generate FM-UWB signal with different characteristics. Measurements with different data rates as well as with different modulation orders are reported in this section. The generator is also capable of producing different scenarios, that include multiple FM-UWB signals in the same RF band, but using different sub-carrier frequencies. The waveforms for different scenarios, as well as the test vectors for the BER measurements are generated using a PC. The same PC is then used to compare the original test vector with the demodulated data recorded by an oscilloscope, and finally produce the BER curves.

The on-chip demodulator performs the first, wideband FM demodulation, and provides the FSK sub-carrier at the output. The second FSK demodulation is performed by software. The receiver output signal is first recorded and digitized using the MSO oscilloscope (that acts as a 10 bit 20 MS/s ADC). The recorded vector is then demodulated using software. The FSK demodulator is implemented as a correlator, which is an optimal maximum-likelihood detector for this case.

As explained previously, two receivers were implemented on the same die in order to compare the receiver performance with the ideal LO and the integrated ring oscillator. Tests were performed using a nominal data rate of 100 kb/s, and the sub-carrier modulation index of 1, meaning that the frequency deviation from center frequency is $\Delta f = 50$ kHz. This is the minimum frequency deviation that preserves orthogonality between the two FSK frequencies for the case of non-coherent signaling. The BFSK sub-carrier signal is not filtered (no pulse shaping is applied), which simplifies the receiver implementation but causes higher ACLR. For this measurement the sub-carrier signal is centered at 1.55 MHz, resulting in two sub-carrier frequencies at 1.5 MHz and 1.6 MHz, although different center frequencies could have been used as well.

In all cases, FSK frequencies are selected so as to have a continuous phase FSK signal. This is generally prefered in order to avoid discontinuities in the signal driving the VCO on the transmitter side, and is therefore used for testing. Sensitivity is defined as the input power that provides a BER of 10^{-3}. The result is shown in Figure 5.18. Measured sensitivity of the receiver with an external LO signal is around -72 dBm. The approximate calculation presented by Gerrits in [1] suggests a sensitivity of -78 dBm for the noise figure of 18 dB. The difference is a result of imperfections present in the implemented receiver, most likely lower gain and increased noise figure of the RF frontend compared to the values obtained by simulation. The measured sensitivity with the internal ring oscillator is -70 dBm, a value approximately 2 dB worse than the sensitivity of the receiver with the external LO. This difference is a result of several factors. First, due to the limited frequency resolution of the internal DCO, it can never be configured to generate the carrier at exactly 4 GHz, meaning that a slight frequency offset will always be present. In this case, the minimum offset that could be achieved was 10 MHz. The second factor that deteriorates the sensitivity is the phase noise of the ring that after demodulation translates into the amplitude noise of the sub-carrier signal and degrades the SNR. The third factor is the amplitude of the LO signal. In the case of an external LO it was increased to provide the best achievable performance. The amplitude was set to 600 mV peak-to-peak at the receiver LO inputs, which results in approximately 420 mV after the RC-CR circuit. This value is larger than the simulated 300 mV peak-to-peak amplitude, that could be generated by the internal LO. The resulting difference is a small price to pay in order to reduce the power of the DCO, although it still remains the most power-hungry block in the receiver.

Additional degradation of sensitivity is expected as the frequency offset increases, as depicted in Figure 5.18(b). Each BER curve was measured after incrementing the DCO control word, roughly corresponding to 25 MHz increase in frequency offset. As shown in the previous chapter, the demodulator conversion gain decreases with the increase of the frequency offset. This effect is further emphasized by the finite bandwidth of the IF amplifier that attenuates the signal amplitude at the edges of the band for a frequency offset above 50 MHz. The effect can be observed in the output waveforms, shown in

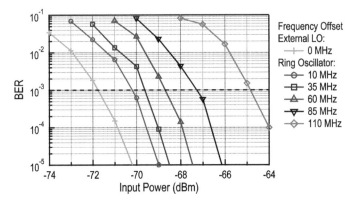

(a) BER curves for different carrier offset

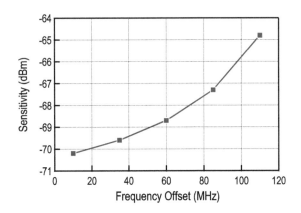

(b) Sensitivity vs. carrier frequency offset

Figure 5.18   Measured BER curves for different carrier offset.

Figure 5.19. As the frequency offset increases the demodulated signal further deviates from the sine wave, thus increasing the power contained in the higher harmonics. The final result is the sensitivity degradation of about 5 dB for the frequency offset of 110 MHz. Depending on the maximum sensitivity degradation that can be allowed, the maximum tolerable frequency offset can be defined, which then translates into the maximum period between the two calibrations and the power overhead due to calibration [15].

Although different receiver architectures have been explored, most of them focused only on standard 2-FSK sub-carrier modulation, targeting data

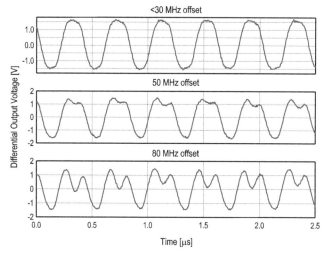

Figure 5.19 Measured demodulator output waveform for different carrier frequency offsets.

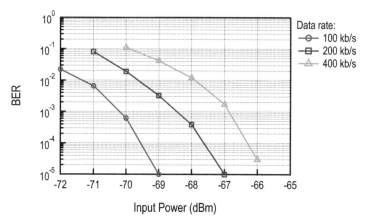

Figure 5.20 BER curves for different data rates.

rates of 100 kb/s and below. Transmitters proposing higher data rates and higher order M-FSK modulations have been implemented, but the full communication with one of the existing receivers has never been demonstrated. In principle, any kind of modulation can be combined with wideband FM modulation to produce the FM-UWB signal. The proposed receiver can then be used to perform the first FM demodulation, while the subsequent sub-carrier demodulation is performed digitally, by software. Two cases are

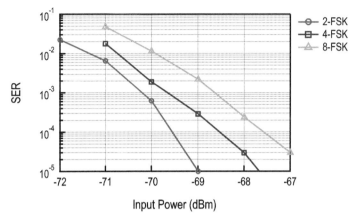

Figure 5.21   BER curves for different modulation order.

of interest here, first is increasing data rate, and second is increasing the modulation order.

The BER curves for different data rates are shown in Figure 5.21. In all cases orthogonal, continuous phase FSK modulation is used. Modulation index is kept constant at 1, meaning that the frequency deviation and the sub-carrier bandwidth increase proportionally to data rate. The limit for the implemented receiver is coming from the bandwidth of the LF amplifier that was intended for operation from 1 MHz to 2.2 MHz. It could easily be extended, at an almost negligible increase in power consumption, if higher data rates are needed. In this case, the receiver is tested up to 400 kb/s. As expected, the sensitivity degrades as the data rate increases, but at a slower rate than in the case of narrow-band modulations, where doubling the data rate results in sensitivity shift of 3 dB. This is a consequence of the non-linear wideband FM demodulator characteristic and is typical for FM-UWB.

Measurements in Figure 5.21 show symbol error rate (SER) results for different FSK modulation orders. The equivalent BER depends on the used coding scheme, and should always be better than the SER. Assuming the same equivalent data rate, increasing modulation order leads to better performance in terms of equivalent BER. This comes at a price of increased sub-carrier signal bandwidth, and demodulator complexity that grows exponentially with the number of bits per symbol. In the reported measurements, the symbol rate is kept constant at 100 ksym/s, leading to 200 kb/s for 4-FSK and 300 kb/s for 8-FSK modulation. Figure 5.22 shows the sub-carrier spectrum at the output of the FM demodulator, shape and bandwidth depend

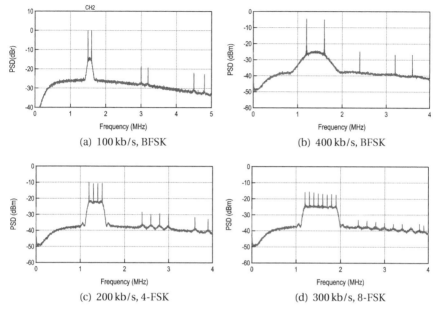

Figure 5.22 Spectrum of the demodulated sub-carrier signal.

on the modulation order and data-rate. In principle, the proposed FM-UWB receiver could use different modulations and data rates to conform to channel conditions (e.g. if path loss is low, higher throughput can be achieved) and available sub-carrier bandwidth, and optimize network performance.

The FM-UWB modulation scheme inherently provides some robustness against narrow-band interferers. In the process of demodulation, the interferer itself is transformed into a dc component that can be filtered out. The cross product of the interferer and the FM-UWB signal results in a component that is spread over a large bandwidth and effectively increases the noise floor at the receiver output [1]. The performance of this receiver in the presence of a narrow-band interferer is shown in Figure 5.23. Sensitivity slowly degrades with the increase of interferer power up to -48 dBm. After that point, the interference becomes the dominant factor that causes erroneous reception and sensitivity begins to degrade linearly with the interference power. The lowest SIR that can be tolerated by this receiver is -17 dB (at -48 dBm interferer power) for an interferer frequency offset of 100 MHz from the center frequency.

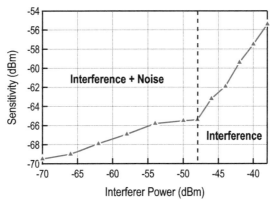

Figure 5.23   Sensitivity as a function of in-band interferer power.

5.4.3 Multi-User Measurements

As shown previously, multiple FM-UWB signals transmitted in the same RF band can be demodulated simultaneously as long as the sub-carrier frequencies are different. At the demodulator output, the FSK modulated useful components will be found along with the spread component that is a result of the cross product between two or more FM-UWB signals. Unlike the useful components that can be filtered out in the baseband, the spread component will always be present and will cause the degradation of sensitivity as the power or number of interfering FM-UWB signals increase. The BER curves in the case of two FM-UWB users are presented in Figure 5.24. The measured sub-carrier channel is the same as in the single user case, centered at 1.55 MHz. Same parameters were used for both channels, 100 kb/s data rate and modulation index of 1, corresponding to sub-carrier bandwidth of 200 kHz. The interfering channel is centered at 1.25 MHz, which provides 100 kHz spacing between the two sub-carrier channels to avoid excessive ACPR. As expected, sensitivity decreases with the increasing power of the interfering FM-UWB signal. The quadratic characteristic of the demodulator will cause the sensitivity degradation to occur quite rapidly. As an example, a 3 dB stronger interferer at the RF input results in 6 dB stronger FSK sub-carrier in the baseband. In order to tolerate significant difference of power levels, high dynamic range baseband circuitry would be needed together with sharp channel filtering. In this case, channel filtering is performed in the digital domain, using a band-pass FIR filter. Since the interfering signals are not filtered before the analog to digital conversion, receiver dynamic range is limited by the dynamic range of the output buffer. Instead of increasing

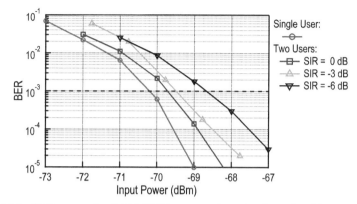

Figure 5.24 BER curves for 2 FM-UWB users and varying input level between the two users.

the dynamic range of the receiver, which would inevitably result in increased power consumption, the problem could be approached at the protocol level by regulating the power on the transmitter side. For the case of HD-WSNs, this approach should not carry too much overhead in terms of complexity as the nodes should not move significantly relative to each other (in comparison with, for example, CDMA in the cellular network).

Just like the inter-user interference increases with the increasing power of the second FM-UWB signal, the increasing number of FM-UWB users will also increase the inter-user interference [16]. Figure 5.25 shows the scenario where the number of users increases from 1 to 4, while the power remains equal in all the channels. The increasing number of channels leads to degraded sensitivity and, just as in the previous case, requires larger dynamic range. Limiting factors to the number of channels, are sub-carrier frequencies, ACLR, channel separation, dynamic range, data rate and inter-user interference. For the given system parameters, 200 kHz wide channels with 100 kHz channel spacing, a maximum of four channels can be used if the lowest sub-carrier channel is located at 1.25 MHz. The demodulator output spectrum is presented in Figure 5.26 for different multi-user scenarios. The measured channel (channel 2, centered at 1.55 MHz) is highlighted in green, and the interfering channels are highlighted in red. A different number of occupied channels can be observed in different figures. As suspected, it can be seen that the spectrum above 2.3 MHz is polluted by the harmonics of the sub-carrier signals and intermodulation products, thus preventing the placement of additional channels in this band. The harmonics

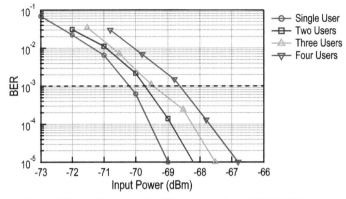

Figure 5.25   BER curves for different number of FM-UWB users.

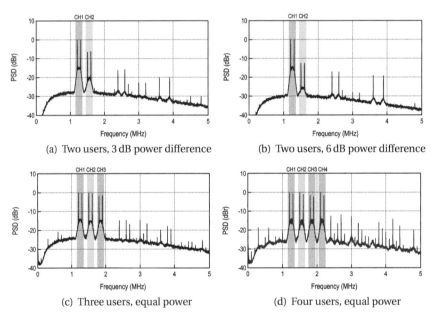

(a) Two users, 3 dB power difference

(b) Two users, 6 dB power difference

(c) Three users, equal power

(d) Four users, equal power

Figure 5.26   Spectrum of the demodulated sub-carrier signal, in different multi-user scenarios.

that can be observed in Figure 5.26 are a combined result of the IF amplifier bandwidth, demodulator bandwidth and the non-linearity of the output buffer.

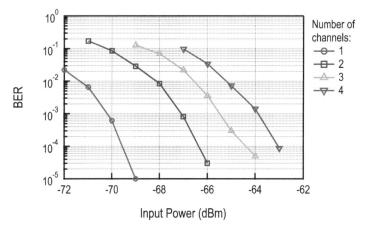

Figure 5.27 BER curves for different number of broadcast sub-channels.

5.4.4 Multi-Channel Transmission Measurements

The multi-channel (MC) transmission concept was also tested with the implemented receiver. The BER curves for different number of sub-channels are shown in Figure 5.27. Compared to the case with multiple transmitters, a significant sensitivity loss can be observed. This is a consequence of scaling since the SNR of a single channel is proportional to $1/M^2$, where M is the number of sub-channels. Nevertheless, for short range applications, where distance between nodes does not exceed several meters, such as BAN, the proposed scheme could still be used and could be of particular interest when a large number of nodes are present and need to receive different data simultaneously. The transmitted spectrum is compared in Figure 5.28 for the cases of a single FM-UWB signal and multiple sub-carrier FM-UWB signal. Since the modulating signal is no longer a triangular waveform, spectral flatness is lost.

One important difference between SC-FDMA with multiple transmitters and a single transmitter is that, in the latter, sub-channels are perfectly synchronized. For multiple transmitters, even if the orthogonal frequencies are used for different sub-channels, the orthogonality is preserved only if symbols are perfectly synchronized. Since it is practically impossible to synchronize multiple transmitters, sub-channels must be separated and a channel filter is required. In the case of a single transmitter, the sub-channels remain perfectly orthogonal, so there is no need for separation, and hence more channels can be placed in the same sub-carrier band. As long as the orthogonality is

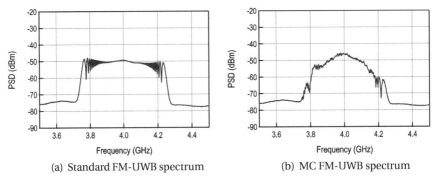

(a) Standard FM-UWB spectrum   (b) MC FM-UWB spectrum

Figure 5.28   Spectrum of the transmitted signal, for the standard FM-UWB and MC FM-UWB.

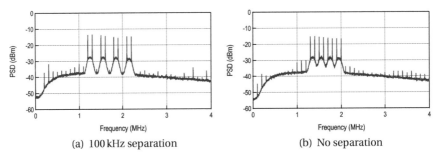

(a) 100 kHz separation   (b) No separation

Figure 5.29   Demodulated signal spectrum, with and without spacing between adjacent sub-channels.

maintained, this separation will not influence the BER. In this regard, the proposed FDMA scheme is similar to the Orthogonal Frequency Division Multiplexing (OFDM) combined with the FM-UWB spread spectrum technique. The spectrum of the demodulated sub-carrier signal with and without separation is shown in Figure 5.29. Measured BER was not influenced by the channel separation.

5.5 Summary

The proposed receiver is compared to the State-of-the-Art receivers in Table 5.2. Of all the implemented FM-UWB receivers it consumes the lowest amount of power while still attaining sufficient sensitivity for short range communications in a HD-WSN. The delay-line demodulator based receivers from [12, 13] have an order of magnitude higher power consumption. The

Table 5.2 Comparison with the state-of-the-art receivers

Parameter	[13]	[12]	[18]	[19]	[17]	This Work
SC-FDMA	**Yes**	**Yes**	–	**No**	**No**	**Yes**
Demodulator	DL	DL	Regen.	Regen.	Regen.	DL
Frequency	7.5 GHz	4 GHz	3.75 GHz	8 GHz	4 GHz	4 GHz
Power Conns.	**9.1 mW**	**10 mW**	**3.8 mW**	**0.6 mW**	**580 μW**	**423 μW**
Supply	1.8 V	2.5 V	1 V	1 V	1 V	1 V
Max. Data Rate	50 kb/s	62.5 kb/s	100 kb/s	1 Mb/s	100 kb/s	400 kb/s
Sensitivity	−88 dBm	−46 dBm	−78 dBm	−76 dBm	−80.5 dBm	−70 dBm
Efficiency	**182 nJ/b**	**160 nJ/b**	**38 nJ/b**	**0.6 nJ/b**	**5.8 nJ/b**	**1.06 nJ/b**
Technology	0.25 μm BiCMOS	0.18 μm BiCMOS	65 nm CMOS	65 nm CMOS	65 nm CMOS	65 nm CMOS

receiver in [17] achieves comparable consumption while providing better sensitivity. The low power consumption is obtained by using a narrow-band amplifier at the input. Since the demodulation is performed using a high-Q RF filter, with a very non-linear FM-AM conversion characteristic, it will not be possible to distinguish between different FM-UWB users. A modification of a regenerative receiver was proposed in [18] that uses two RF filter paths to achieve better linearity and loosen the Q constraints, but it consumes 3.8 mW. This receiver could potentially be utilized in a multi-user scenario; however, this capability was not demonstrated. The same receiver architecture was used to demodulate a Chirp-UWB signal in [19]. Even though the receiver consumes a peak power of 4 mW, the average power is decreased to 0.6 mW by employing duty cycling. The receiver proposed here already consumes low power in continuous operation, however, by applying the same duty-cycling technique, its power consumption could be reduced below 200 μW, which might be addressed in future research.

Aside from the low power achieved, the proposed receiver offers the capability for several FM-UWB users to communicate in the same RF band at the same time. In an environment where a lot of nodes need to operate in a small area, SC-FDMA can provide more flexibility for protocol optimization, and lead to lower latency by allowing multiple nodes to communicate at the same time. The only two other receiver implementations offering the same capability require an order of magnitude higher power, thus making the proposed receiver a better solution for the given scenario. In addition, the implemented receiver could support different data rates and different M-FSK modulations. With a flexible digital baseband it would be possible to dynamically adjust the number of channels and data rate per channel, allowing to further optimize network performance.

References

[1] J. F. M. Gerrits, M. H. L. Kouwenhoven, P. R. van der Meer, J. R. Farserotu, and J. R. Long, "Principles and limitations of ultra-wideband FM communications systems," *EURASIP J. Appl. Signal Process.*, vol. 2005, pp. 382–396, Jan. 2005.

[2] A. Shameli and P. Heydari, "A novel ultra-low power (ULP) low noise amplifier using differential inductor feedback," in *2006 Proceedings of the 32nd European Solid-State Circuits Conference*, Sept. 2006, pp. 352–355.

[3] A. C. Heiberg, T. W. Brown, T. S. Fiez, and K. Mayaram, "A 250 mv, 352 μW GPS receiver RF front-end in 130 nm CMOS," *IEEE Journal of Solid-State Circuits*, vol. 46, no. 4, pp. 938–949, April 2011.

[4] T. Taris, J. Begueret, and Y. Deval, "A 60 μW lna for 2.4 GHz wireless sensors network applications," in *2011 IEEE Radio Frequency Integrated Circuits Symposium*, June 2011, pp. 1–4.

[5] E. M. Cherry and D. E. Hooper, "The design of wide-band transistor feedback amplifiers," *Electrical Engineers, Proceedings of the Institution of*, vol. 110, no. 2, pp. 375–389, Feb. 1963.

[6] C. Holdenried, J. Haslett, and M. Lynch, "Analysis and design of HBT Cherry-Hooper amplifiers with emitter-follower feedback for optical communications," *IEEE Journal of Solid-State Circuits*, vol. 39, no. 11, pp. 1959–1967, Nov. 2004.

[7] C. D. Holdenried, M. W. Lynch, and J. W. Haslett, "Modified CMOS Cherry-Hooper amplifiers with source follower feedback in 0.35 μm technology," in *Solid-State Circuits Conference, 2003. ESSCIRC '03. Proceedings of the 29th European*, Sept. 2003, pp. 553–556.

[8] N. Saputra and J. Long, "A Fully-Integrated, Short-Range, Low Data Rate FM-UWB Transmitter in 90 nm CMOS," *IEEE Journal of Solid-State Circuits*, vol. 46, no. 7, pp. 1627–1635, July 2011.

[9] M. Grozing, B. Phillip, and M. Berroth, "CMOS ring oscillator with quadrature outputs and 100 MHz to 3.5 GHz tuning range," in *Solid-State Circuits Conference, 2003. ESSCIRC '03. Proceedings of the 29th European*, Sept. 2003, pp. 679–682.

[10] E. J. Pankratz and E. Sanchez-Sinencio, "Multiloop high-power-supply-rejection quadrature ring oscillator," *IEEE Journal of Solid-State Circuits*, vol. 47, no. 9, pp. 2033–2048, Sept. 2012.

[11] J. Chabloz, D. Ruffieux, and C. Enz, "A low-power programmable dynamic frequency divider," in *Solid-State Circuits Conference, 2008. ESSCIRC 2008. 34th European*, Sept. 2008, pp. 370–373.

[12] J. Gerrits, J. Farserotu, and J. Long, "A wideband FM demodulator for a low-complexity FM-UWB receiver," in *The 9th European Conference on Wireless Technology, 2006*, Sept. 2006, pp. 99–102.

[13] Y. Zhao, Y. Dong, J. F. M. Gerrits, G. van Veenendaal, J. Long, and J. Farserotu, "A short range, low data rate, 7.2 GHz-7.7 GHz FM-UWB receiver front-end," *IEEE Journal of Solid-State Circuits*, vol. 44, no. 7, pp. 1872–1882, July 2009.

[14] E. A. Vittoz and O. Neyroud, "A low-voltage cmos bandgap reference," *IEEE Journal of Solid-State Circuits*, vol. 14, no. 3, pp. 573–579, June 1979.

[15] N. Pletcher, S. Gambini, and J. Rabaey, "A 52 μW wake-up receiver with 72 dBm sensitivity using an Uncertain-IF architecture," *IEEE Journal of Solid-State Circuits*, vol. 44, no. 1, pp. 269–280, Jan. 2009.

[16] J. M. F. Gerrits, J. R. Farserotu, and J. R. Long, "Multi-user capabilities of UWBFM communications systems," in *2005 IEEE International Conference on Ultra-Wideband*, Sept. 2005.

[17] N. Saputra, J. Long, and J. Pekarik, "A low-power digitally controlled wideband FM transceiver," in *2014 IEEE Radio Frequency Integrated Circuits Symposium*, June 2014, pp. 21–24.

[18] F. Chen, W. Zhang, W. Rhee, J. Kim, D. Kim, and Z. Wang, "A 3.8-mW 3.5-4-GHz regenerative FM-UWB receiver with enhanced linearity by utilizing a wideband LNA and dual bandpass filters," *IEEE Transactions on Microwave Theory and Techniques*, vol. 61, no. 9, pp. 3350–3359, Sept. 2013.

[19] F. Chen, Y. Li, D. Liu, W. Rhee, J. Kim, D. Kim, and Z. Wang, "A 1 mW 1 Mb/s 7.75-to-8.25 GHz chirp-UWB transceiver with low peak-power transmission and fast synchronization capability," in *2014 IEEE International Solid-State Circuits Conference Digest of Technical Papers (ISSCC)*, Feb. 2014, pp. 162–163.

[20] J. Farserotu, J. Baborowski, J. D. Decotignie, P. Dallemagne, C. Enz, F. Sebelius, B. Rosen, C. Antfolk, G. Lundborg, A. Björkman, T. Knieling, and P. Gulde, "Smart skin for tactile prosthetics," in *2012 6th*

International Symposium on Medical Information and Communication Technology (ISMICT), March 2012, pp. 1–8.

[21] Y. Dong, Y. Zhao, J. F. M. Gerrits, G. van Veenendaal, and J. Long, "A 9 mW high band FM-UWB receiver front-end," in *Solid-State Circuits Conference, 2008. ESSCIRC 2008. 34th European*, Sept. 2008, pp. 302–305.

[22] N. Saputra and J. R. Long, "A fully integrated wideband FM transceiver for low data rate autonomous systems," *IEEE Journal of Solid-State Circuits*, vol. 50, no. 5, pp. 1165–1175, May 2015.

[23] ——, "A short-range low data-rate regenerative FM-UWB receiver," *IEEE Transactions on Microwave Theory and Techniques*, vol. 59, no. 4, pp. 1131–1140, April 2011.

[24] V. Kopta, D. Barras, and C. Enz, "A $420\,\mu w$, $4\,GHz$ approximate zero IF FM-UWB receiver for short-range communications," in *2016 IEEE Radio Frequency Integrated Circuits Symposium*, May 2016, pp. 218–221.

[25] V. Kopta, D. Barras, and C. C. Enz, "An approximate zero IF FM-UWB receiver for high density wireless sensor networks," *IEEE Transactions on Microwave Theory and Techniques*, vol. 65, no. 2, pp. 374–385, Feb. 2017.

6

FM-UWB Transceiver

6.1 Introduction

The previous chapter dealt in details with the implementation of the quadrature approximate zero-IF receiver. The next step is to integrate a full FM-UWB transceiver. Aside from the quadrature AZ-IF receiver, that provides the multi-user communication capability, the single-ended receiver is added, and can be used to reduce the power consumption of the transceiver when lower performance is acceptable. Furthermore, the integrated transceiver includes a baseband that performs the sub-carrier FSK demodulation and symbol clock recovery, and provides fully digital outputs that can be interfaced by an FPGA or a microcontroller. Although there are no major architectural innovation on the transmitter side, the clever use of circuit techniques provides means for some improvements compared to the state of the art in terms of power consumption and efficiency. The emphasis is on the fully integrated output matching network, that allows to use the same RF pad for both reception and transmission, eliminating the need for an external switch or any other passive components.

First, the top level architecture of the transceiver is described. Then, in the following section, the details of transmitter circuit implementation are given, followed by the circuit implementation of the two receivers in Section 5.3. The results of transceiver characterization are presented in Section 5.4, demonstrating the capabilities of the proposed approach with emphasis on robustness and low power consumption. Finally, the chapter is concluded with the summary of performance and comparison with the state of the art.

6.2 Transceiver Architecture

The implemented FM-UWB transceiver consists of two receivers and a transmitter (Figure 6.1). Aside from low power consumption, the emphasis of this

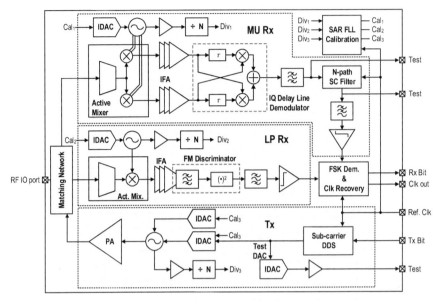

Figure 6.1   Top-level block diagram of implemented transceiver.

work is also on high level of integration and miniaturization, robustness to narrowband interferers and tolerance to reference frequency offset. A single RF pad is used as both receiver input and transmitter output, eliminating the need for an off-chip switch that was needed in all the previous implementations found in the literature. In addition, the matching network is fully integrated and no passive components need to be added externally.

Large signal bandwidth of the FM-UWB signal means that the transceiver inherently possesses some robustness against the RF carrier frequency offsets. In the baseband section of the receiver, a relatively small offset (e.g. less than 0.5%) between the transmit and the receive symbol clock can be compensated using a simple clock recovery scheme. As a result, there is no need for a precise frequency reference in the system and no need for a crystal oscillator. Instead, an RC reference oscillator can be integrated on-chip and calibrated before use, allowing to completely remove all the external components.

Just like in the previous section, the LO signal is generated using ring oscillators. Owing to the loose phase noise constraints and the large signal bandwidth, such an approach is acceptable and is exploited to reduce the receiver power consumption. Normally, ring oscillators operate in a loop that stabilizes the oscillation frequency (PLL), however here they are used in a free-running mode so as to reduce the significant power overhead that would otherwise be present due to continuous operation of frequency dividers. Instead, the oscillators are periodically calibrated using a successive approximation register (SAR) FLL, assuring that the frequency offset remains within the required limits.

The transmitter architecture is similar to other implementations found in the literature. The sub-carrier signal is synthesized digitally, allowing easier control and switching of sub-carrier frequencies, as well as better precision compared to an analog solution with a capacitor bank. The sub-carrier signal is used to drive a current DAC that controls the DCO frequency, that finally produces the desired FM-UWB signal. The resulting signal is then amplified by the preamplifier (PPA) and the power amplifier (PA) before transmission.

The two implemented receivers are intended for two modes of operation. The multi-user (MU) receiver consumes more power and is capable of distinguishing multiple FM-UWB signals and providing SC-FDMA. Its purpose is to provide multiple channels and speed up communication when network traffic is high. It is based on the receiver described in the previous section with the addition of the channel filtering and baseband processing. The low power (LP) receiver provides a low power, low performance mode, that can be used when the network traffic is low and a single channel is sufficient. Instead of quadrature demodulation, it only uses a single branch, allowing to simplify the receiver architecture and save power. However, due to the non-linearity of the frequency-to-amplitude conversion characteristic of the demodulator, it cannot distinguish multiple channels allowing only a single FM-UWB user.

6.3 Transmitter Implementation

A more detailed block diagram of the transmitter is given in Figure 6.2. The DCO is in fact driven by two DACs. The first one, referred to as the static DAC, only determines the highest frequency in the FM-UWB signal spectrum f_H. Its output remains constant during transmission. The second DAC, or the dynamic DAC, is driven by the digital sub-carrier signal and controls instantaneous frequency of the DCO, and hence all the modulation characteristics. Two DACs are calibrated prior to transmission using an on-chip SAR FLL. In the first step, the static DAC is calibrated to set f_H. In the second step, the dynamic DAC is calibrated to set the FM-UWB signal bandwidth. The third DAC is added for testing purposes. It is an exact copy of the dynamic DAC that provides the sub-carrier signal at its output, allowing to verify the correct operation of the SC-DDS and to measure the sub-carrier frequency.

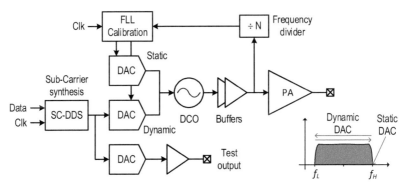

Figure 6.2   Block diagram of the implemented transmitter.

The DCO is separated from the PA and frequency dividers by buffers. These buffers prevent frequency pulling that might be caused by an outside signal, or by the change in capacitance value in the on and off state of the frequency divider. The frequency divider is implemented in the same way as in the receiver described in the previous chapter. It is a chain of ten divide-by-2 circuits from [1] that provide the signal for calibration. The divider buffer, that is needed to amplify the DCO signal and provide the rail-to-rail swing at its output, is adjusted here for a single-ended input.

6.3.1 Sub-Carrier Synthesis

The sub-carrier synthesizer is a fully digital block that provides a 6 bit value at its output. This implementation is following the principles described in [2]. The sub-carrier signal is generated in two steps. First, an accumulator generates a saw-tooth waveform, and then in the second step this saw-tooth waveform is folded to produce a triangular waveform. The frequency of the sub-carrier is controlled by controlling the slope of the saw-tooth waveform, or equivalently the increment value of the accumulator (SC1 and SC2 in Figure 6.3). The sub-carrier frequency is given by

$$f_{sc} = f_{clk} \frac{M}{2^N},$$

(6.1)

where f_{clk} is the input clock frequency, N is the number of bits of the accumulator, and M is the increment. Although only 6 bits are used to control the DAC, 16 bits are used for the accumulator to provide the needed frequency resolution. Nominal clock frequency is 40 MHz, which results in frequency resolution of approximately 600 Hz, which is more than enough

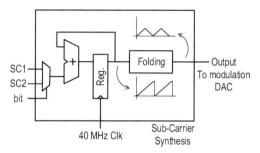

Figure 6.3 Digital sub-carrier synthesizer.

to generate a 100 kb/s FSK signal with a modulation index of 1. No pulse shaping or filtering is used in this implementation. Compared to fully analog implementations, such as those described in [3, 4], DDS approach requires slightly more power. However, since the SC synthesizer consumption is typically small compared to other blocks in the transmitter, this overhead is negligible, and DDS provides better frequency precision (relative to the reference clock) and easier control, both highly desirable especially in a multi-user scenario.

6.3.2 DCO Digital to Analog Converters

Two current mode digital to analog converters are used to drive the current starved ring oscillator, as explained previously. The static DAC is used to set the high frequency of the FM-UWB signal f_H and the dynamic DAC is used to generate the FM-UWB modulation.

The static DAC is shown in Figure 6.4. Digital control word can be either written manually, using the SPI bus, or a calibration loop can be used to set the register value. Once set, the control word remains constant throughout the transmission. Therefore, there are no specific constraints regarding speed or glitches, and a relatively simple solution can be used. Six bits $(b_0 - b_5)$ control the binary weighted current mirror that provides the bias current of the ring oscillator. Although linearity is not paramount for the static DAC, a relatively good characteristic is obtained (as shown in Figure 6.9). The only requirement of this DAC is monotonicity of the characteristic, that is needed to assure the proper functionality of the SAR FLL calibration scheme. Considering that only 6 bits are used, no special matching techniques are necessary to achieve the desired precision.

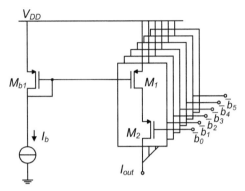

Figure 6.4   Static DCO current DAC.

The dynamic DAC is shown in Figure 6.5. This DAC is actually sinking current from the static DAC. As the value of the digital control word increases, so does the current, effectively reducing the bias current of the oscillator and consequently its frequency. This approach keeps current of the static source constant and provides the good linearity needed for FM-UWB signal generation. As opposed to the static DAC, cascode current mirror is used here, in order to provide better precision of the output current. The dynamic DAC is controlled by the digital sub-carrier signal, which means that it must operate at the clock frequency of 40 MHz, even though the sub-carrier frequency is below 2.5 MHz. For this reason, the current steering approach is used. In this way, the bottom reference transistors M_3 and M_4 never switch off, and there is no additional delay coming from the time required for transistors to turn on (time it takes to charge gate capacitance through the bias transistors M_{b1} and M_{b2}). As a consequence, this technique produces smaller glitches at the output. A capacitor is placed at the output of the two DACs to filter the DCO current and avoid sharp pulses in DCO supply current.

Bandwidth of the FM-UWB signal is controlled using the reference current I_b. The sub-carrier DDS signal always produces a full scale DAC output (all 6 bits are used), and the reference current determines the maximum value of the current, and subsequently the lower frequency f_L. The reference current is generated using yet another current DAC (again static during transmission), with a 5 bit resolution. To calibrate the reference current, bits b_0 to b_5 are all set to '1', resulting in the maximum output current, and then the SAR FLL sets the control bits of the reference current DAC to provide the desired lower frequency f_L. In this way, FM-UWB bandwidth is entirely decoupled from the sub-carrier generation.

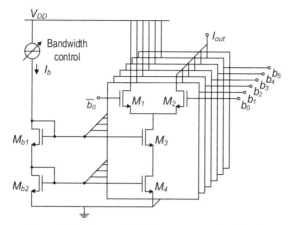

Figure 6.5 Dynamic DCO current steering DAC.

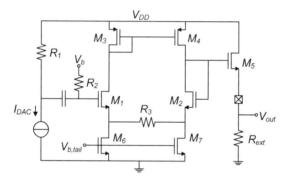

Figure 6.6 Dynamic DAC test output buffer.

As explained above, an exact replica of the dynamic DAC is added for testing. This DAC is driven from the same SC DDS circuit, and will add additional capacitive load to its output. However, since the consumption of the SC DDS remains negligible compared to the DCO and the PA, presence of the testing circuits will not have a significant impact on the overall power consumption. The test DAC drives a buffer that provides analog sub-carrier signal at its output. Resistor R_1 converts the DAC output current into the input voltage of the buffer. This buffer is shown in Figure 6.6. Its main purpose is to verify the sub-carrier frequency. It consists of a resistively degenerated differential pair and a source follower that provides a low impedance output capable of driving a 10 pF capacitive load. Bias current and bandwidth of the source follower are set by the external resistor R_{ext}.

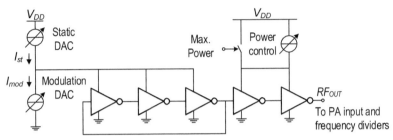

Figure 6.7   Transmitter DCO with buffers.

6.3.3 DCO

The implemented transmitter DCO is shown in Figure 6.7. Since only a single-ended output is needed, the simplest topology, that consequently consumes the lowest amount of power for a given frequency of oscillation, is used. The three inverters of the ring oscillator are scaled progressively, with increasing transistor width from left to right (for both NMOS and PMOS). This was done to increase the driving capability of the inverter driving the buffer, without increasing the overall power consumption. A ring oscillator that is not fully symmetric, i.e. with different stages, generally exhibits higher phase noise [5]. However, in this case symmetry is already broken by the presence of the buffer, and phase noise is not a limiting factor due to the large signal bandwidth, allowing to concentrate on power reduction.

Buffers are placed between the DCO on one side and preamplifier and frequency divider driver on the other. Since the input capacitance of the driver changes in on and off state, buffer is needed to avoid frequency shift after calibration. In addition, it provides isolation in order to avoid frequency pulling if a strong external signal is present. The buffer inverters are current starved, allowing to control the output amplitude by controlling the supply current. There is also a possibility to bypass the current source and connect the buffers directly to the supply voltage, providing the maximum amplitude. Driver of the frequency divider (shown in Figure 6.8) is needed to amplify the DCO output and provide a rail-to-rail signal. It is designed to provide a rail-to-rail signal for a minimum input amplitude of 10 mV, at 5.5 GHz. It consists of 5 inverter stages, two of which are self biased using a large resistor. Since the driver will be on only when calibration is needed it will not contribute significantly to the overall power consumption.

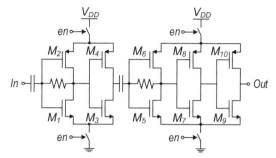

Figure 6.8 Schematic of the frequency divider buffer.

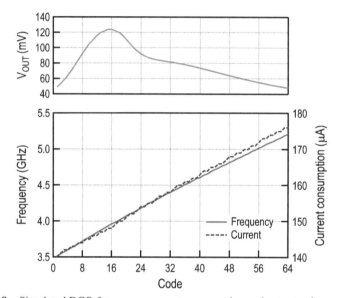

Figure 6.9 Simulated DCO frequency, current consumption and output voltage amplitude.

Figure 6.9 shows simulated frequency, current consumption and output amplitude of the designed DCO, including the two buffer inverters. Resonant load at the preamplifier input is used to boost the signal amplitude close to 4 GHz. The DCO covers the range from 3.5 GHz to 5.2 GHz, and with the 6 bit DAC, this results in frequency resolution of roughly 25 MHz. At 4 GHz, the DCO consumes 150 μW, for a 120 mV signal amplitude at the output. The linearity of the DCO is sufficient for the generation of the FM-UWB signal, and does not cause notable distortion in the output spectrum.

6.3.4 Preamplifier and Power Amplifier

The output stage of the transmitter consumes the largest portion of power and is therefore the most critical for the overall performance. As any other FM modulation, the FM-UWB is a constant envelope modulation, meaning that there is no need for use of special techniques such as outphasing, envelope elimination and restoration, adaptive biasing etc. However, low constraint in terms of output power (maximum output power is below -10 dBm), combined with large bandwidth of the FM-UWB signal, result in more complex output matching network, and consequently lower achievable efficiency.

Design of a power amplifier for low output power poses specific challenges, usually different than those seen in more common applications. Generally, for every power amplifier there is an optimal load impedance that results in highest efficiency for a given output power. Consider a PA that needs to provide 20 dBm output power. For a 50 Ω load this translates into a voltage swing of more than 6 V peak to peak. This becomes difficult to achieve in deep sub-micron technologies with a low supply voltage, and a matching network is required that will lower the load impedance to the optimal value, much smaller than 50 Ω. For the case of FM-UWB, the maximum output power is limited to -10 dBm, which translates into a 200 mV swing over a 50 Ω load. In this case, the optimal impedance seen from the PA needs to be higher than the load, and the matching network instead needs to boost the load impedance to maximize efficiency. This adds different constraints and changes the design approach compared to the first case. The second important difference between the high and low power PA design is in the driving circuits. At 20 dBm output power, driving circuits will consume only a small portion of the overall power, and will not affect the efficiency significantly. They become much more important when the output power becomes comparable to the consumption of the driving circuit and may greatly affect the choice of the PA class of operation.

In most mid and high power applications, with a constant envelope modulation, class E PA is commonly used as the most efficient solution. To achieve high efficiency, such PA requires a matching network precisely tuned to a certain frequency. Achieving high efficiency over a wide bandwidth (500 MHz in this case) becomes a difficult task with class E. Additionally, class E and other switching PAs have very high driving requirements, which pose problems for the driving circuits at low output power levels. For proper class E operation the switch needs to be driven with a full swing rectangular signal, with very short transition times, in order to minimize switching losses.

In addition, the efficiency of every switching PA is inversely proportional to the switch on resistance. This condition sets a limit to the minimum size of the PA transistor and consequently sets its input capacitance. Since the driver consumption is proportional to fCV_{DD}^2, at 4 GHz and -10 dBm output power, it will become comparable to the PA consumption, resulting in significant overall efficiency penalty. For this reason, linear power amplifiers are a better choice for this application.

For the proper choice of a linear PA, its efficiency must be taken into account together with the needed input signal amplitude. Going from class A to class C, the conduction angle of the PA decreases, and the efficiency increases. However, to maintain the same output power, the input signal must increase its amplitude at lower conduction angles, thus imposing higher driving requirements. Higher input amplitude means higher preamplifier consumption (proportional to the square of the amplitude). A good trade-off between driving requirements and PA efficiency is a class AB amplifier and is therefore used in this design.

As explained previously, for -10 dBm output power and 1 V supply, optimal impedance seen from the PA is proportional to V_{DD}^2/P_{out}, which is in this case much larger than 50 Ω [6, 7]. The matching network that will implement this ratio is difficult to implement on chip due to the limited quality factor and limited inductance value of the integrated inductors. The problem can be solved by lowering the supply voltage which consequently lowers the optimal impedance and the transformation ratio. Instead of using a separate circuit to lower the supply voltage, such as a DC-DC converter, the preamplifier and the PA can be stacked (similarly to the approach from [3]). At the same time, this simplifies the matching network, as the equivalent PA supply is reduced, and saves power since the preamplifier reuses the bias current of the PA.

The designed output stage, with the preamplifier and the main PA is shown in Figure 6.10. Capacitors C_{in1} and C_{in2}, together with the inductor L_{PPA} provide the resonant load for the DCO buffer in order to boost the amplitude around 4 GHz. Capacitor C_{in2} can be tuned to compensate for process variations. Bias current of the entire stage is determined by the bias point of the PPA transistor M_1. Just like the main PA, the PPA is also biased in class AB. Output power can be controlled by controlling the bias point of M_1. Filter that consists of the RF choke L_{PPA} and a decoupling capacitor C_{dec} provides a steady voltage at the source of M_2. Depending on the bias current this voltage will vary from 0.3 V to 0.4 V, and is determined by the V_{GS} of M_2 and M_3. The resonance frequency at the PPA output is determined by the L_{PPA} and the equivalent capacitance seen from the drain of M_1, which

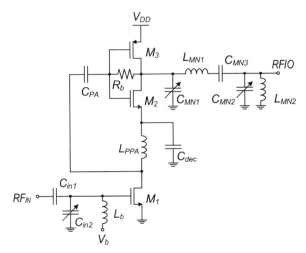

Figure 6.10   Preamplifier and power amplifier schematic.

is mainly determined by the C_{PA} and the gate capacitances of M_2 and M_3. The two resonant frequencies at the input and output of the PPA are offset from 4 GHz in opposite directions in order to provide a relatively constant signal amplitude at the PA input over 500 MHz bandwidth. The main PA is a complementary class AB power amplifier. It is self biased via a 95 kΩ resistor R_b. The PA input amplitude above 250 mV in the desired band is enough to drive the class AB amplifier in saturation and provide a relatively good efficiency. Output matching network that consists of inductors $L_{MN1,2}$, and capacitors C_{MN1-3} transforms the output 50 Ω impedance into roughly 700 Ω over the entire band of operation, as seen from the PA output. This is enough to provide an almost rail to rail signal at the PA output (with respect to the PA supply, meaning from 0.3 V to 1 V), that minimizes power dissipation in the two output transistors.

The output matching network is designed together with the matching network of the two LNAs. To simplify the overall design procedure, the LNA is designed to present a capacitive load at the RF IO in the off state. The total capacitive load at the RF IO is determined by the pad capacitance and the two LNAs. Since matching requirements are not the same in the receive and transmit mode, the matching network must be able to adjust to both. This can be achieved by adjusting the capacitance of C_{MN1} and C_{MN2}. The value of C_{MN1} changes from 120 fF to 520 fF in transmit and receive mode. The C_{MN2} is actually set to 0 in the transmit mode. Generally, the capacitance at

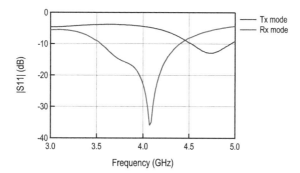

Figure 6.11 Simulated S_{11} parameter at the RF IO.

this node (RF IO) should be minimized in order to maximize the efficiency. Inductor L_{MN2} is placed to partially compensate this capacitance and to extend the bandwidth. In the receive mode, the value of C_{MN2} increases to 600 fF. In the actual implementation 3 bits are used to control the two capacitors, the first bit switches from transmit to receive state, and the additional 2 bits allow some frequency tuning, that allows small modifications of the resonance frequencies once the chip is placed on a PCB. Simulated S_{11} parameter at the RF IO port is shown in Figure 6.11 for the two modes of operation. It can be seen that in the transmit mode, the reflection coefficient is quite high, around -5 dBm in the band of interest, and would not provide adequate matching to a $50\,\Omega$ antenna. Once the capacitors are switched, the input reflection coefficient drops below -10 dBm from 3.6 GHz to 4.35 GHz.

Simulated output power, power consumption and efficiency of the output stage are shown in Figure 6.12. The shown characteristic of the PA is static in the sense that the signal frequency is constant. However, assuming a relatively slow moving FM-UWB carrier, the simulation should provide a good estimate of performance during transmission. The PA was designed to provide roughly constant output power over the desired frequency range of 500 MHz. The low equivalent Q factor of the output matching network results in lower peak efficiency of the amplifier, which is an inherent drawback of wideband power amplifiers, and the price to be paid for large signal bandwidth. As shown in Figure 6.12(a) the simulated output power varies less than 1 dB between 3.7 GHz and 4.4 GHz, with an average output power around -9.4 dBm. This level is somewhat higher than the allowed power in the UWB band. The design is intentionally targeting a higher output power since the actual power level is expected to be lower than the one simulated (due to losses on the PCB and imperfect matching). In the same band power consumption from a 1 V

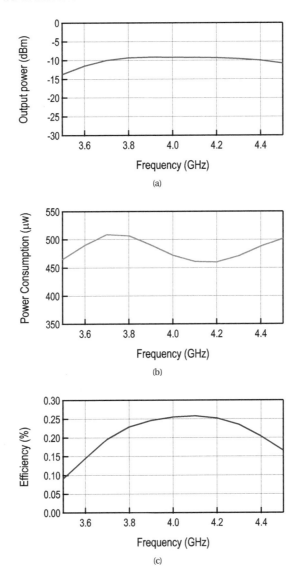

Figure 6.12   Simulated power amplifier output power (a), consumption (b) and efficiency (c) including the preamplifier.

supply varies between $510\,\mu W$ and $460\,\mu W$, with an average value of $481\,\mu W$. The PA efficiency including the PPA is above 20% in the range of interest, with the average simulated efficiency of 24%. With the reported results, the proposed transmitter should consume the lowest amount of power of all the so far implemented FM-UWB transmitters.

6.4 Receiver Implementation

The two implemented receivers, MU and LP receiver will be described in detail here. The MU receiver is very similar to the receiver described in the previous chapter, only minor modifications are done at the circuit level. In the baseband, after the wideband FM demodulator, a channel filter and an FSK demodulator are added, so that the entire processing is now done on-chip. The same FSK demodulator is used by the LP receiver, however, in this case there is no need for the channel filter as only a single FM-UWB channel can be used.

6.4.1 RF Frontend

As in the previous case, the LNA and mixer are stacked (active mixer) to conserve power in both LP and MU receivers. The two schematics of the active mixer for the case of the MU receiver and the LP receiver are shown in Figures 6.13 and 6.14 respectively. Again, a transformer based approach is used to boost the equivalent transconductance of the input transistor to approximately $2G_{m1}$. As opposed to the previous version, there is no complementary input transistor, only the NMOS is used, which simplifies the layout of the LNA and reduces parasitics, however, the output bias point and the LNA gain are no longer decoupled. Switch, sw_1 is placed to disconnect the transformer when transceiver is in the transmit mode. This is done to avoid the impact of the transformer to the PA output impedance and efficiency degradation. Resistors R_{M1-4} are fixed to $15\,k\Omega$, no gain switching is present in this design, although G_{m1} can be varied by slightly tuning the bias current.

The LP RF frontend only has a single differential output, and therefore only needs one LO input. This allows to simplify the oscillator and save power in the the LO generation. Although a single LO signal is present, mixing is done using a current steering differential pair $M_{M1}-M_{M2}$, where gate of M_{M2} is tied to V_{bm}. Because a single-ended LO signal is used, voltage gain will be lower and noise figure will be higher compared to a case with a differential driving signal with the same amplitude. Load of the LP mixer is done using PMOS transistors M_{M3} and M_{M4}, that provide the output bias point and load resistors R_{M1} and R_{M2} that determine the output impedance, allowing some decoupling between the two.

Input matching network is implemented in the same way for both frontends. The input reflection coefficient is given in Figure 6.11, and is the same for both (although there is a small difference in G_{m1}). The MU

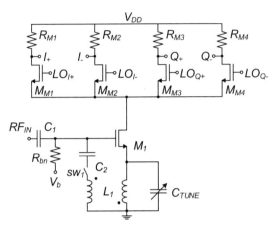

Figure 6.13   MU receiver LNA/mixer schematic.

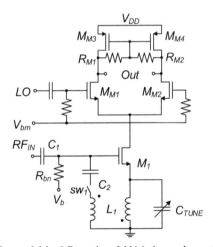

Figure 6.14   LP receiver LNA/mixer schematic.

LNA/mixer provides 13 dB differential voltage gain, together with a noise figure of around 15 dB, while consuming 100 μW. The LP LNA/mixer achieves 11 dB gain and 19 dB noise figure (it should be noted that a single ended LO signal of a lower amplitude is used) while consuming 70 μW. In both MU and LP implementation the main source of noise is transistor M_1. For both active mixers the input referred 1 dB compression point is around $P_{1dB} = -16$ dBm, and the third order intercept point is around $IIP_3 = -3$ dBm.

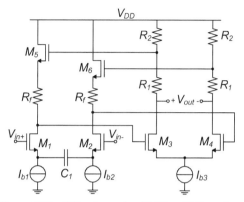

Figure 6.15 IFA schematic of MU and LP receiver.

6.4.2 IF Amplifiers

A Cherry-Hooper amplifier described in the previous chapter was reused in this design. First difference that can be noticed compared to the previous design is that the capacitor C_1, used to prevent offset propagation, is now placed in the source of the first differential pair. The second difference is that resistor R_f is now used to provide gain switching. It was established by simulation that additional parasitics due to switching circuitry at R_f have less impact on the amplifier bandwidth than was the case previously. A single gain control bit is provided per stage, resulting in 3 control bits for the 3 cascaded IF stages, that provide gain switching in roughly 5 dB steps. As in the previous case, lower gain setting slightly extends bandwidth.

Although schematics of the two, LP and MU, IF amplifiers are identical, values of different circuit elements are different in order to conform to slightly different design requirements. Output resistors are slightly lower in the two last stages of the IF amplifier in order to cope with the higher capacitive load at this node. The capacitive load is mostly due to the buffers that precede the wideband FM demodulators. Additional difference exists in the LP IF amplifier and is concerning the capacitor C_1. Namely, in the LP receiver case, the IF amplifier is also part of the FM demodulator, and acts as a frequency discriminator (performs FM-AM conversion). The high pass frequency characteristic needed for demodulation is implemented using a small capacitor C_1. The zero and pole coming from this capacitor are given by

$$z \approx -\frac{1}{2C_1 R_{o,b}}, \quad p \approx -\frac{G_{m1}}{2C_1}, \tag{6.2}$$

where $R_{o,b}$ is the output resistance of the tail current source. The approximation is valid if $G_{m1}R_{o,b} \gg 1$. The cut-off frequency of the high-pass filter is determined by the M_1 transconductance and the source capacitor. This small capacitor of 200 fF is placed at the third stage of the IF amplifier. Large capacitors of 2.5 pF used for the first two stages provide a cut-off frequency below 5 MHz and should not have a significant impact on demodulation. Finally, the chosen values result in a first order high pass characteristic with the cut-off frequency above 200 MHz.

Simulated characteristics of the implemented amplifiers are shown in Figure 6.16. The high-pass characteristic of the IF amplifier in the LP receiver provides the FM-AM conversion before the envelope detector. In the pass band the IF amplifier provides more than 35 dB of gain, which together with the RF frontend results in the conversion gain of around 45 dB. The MU receiver IF amplifier provides around 40 dB gain, resulting in almost 50 dB conversion gain together with the RF frontend. Both amplifiers provide more than 300 MHz bandwidth, that should be enough to compensate for the ±50 MHz carrier frequency offset. Sharp, 6th order filtering characteristic

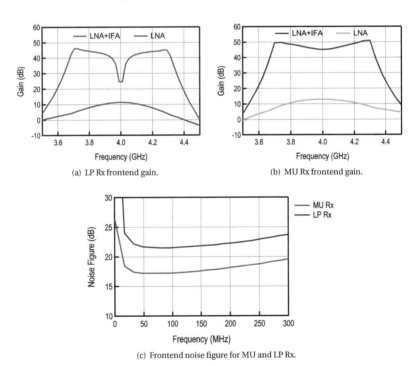

(a) LP Rx frontend gain.

(b) MU Rx frontend gain.

(c) Frontend noise figure for MU and LP Rx.

Figure 6.16   Simulated characteristics of the LP and MU Rx frontend.

provides good rejection of out of band interferers. The total noise figure of all the stages preceding the demodulator is around 22 dB for the LP receiver and around 18 dB for the MU receiver. The limited gain of the LNA/mixer stage, results in increase of the noise figure due to the noise of the IF amplifiers. Higher noise in the LP IF amplifier is coming purely from the RF frontend. Together with the 6 dB difference in sensitivity, coming from the demodulator implementation, the 4 dB difference in noise figure should amount to roughly 10 dB difference in sensitivity between the two receivers. In this case, this sensitivity loss is a price to pay for low power consumption. In both cases, a single IF stage consumes around 20 μW, amounting to a total IF power consumption of 60 μW in the case of LP, and 120 μW in the case of MU receiver.

6.4.3 Receiver DCO

The quadrature DCO described in the previous chapter was reused in the MU receiver without any significant changes. Since the LP receiver does not require quadrature LO signals, a different DCO was designed, allowing to save power needed to generate the LO signal. Knowing that the DCO is one of the biggest consumers in the receiver, such an approach allows to further reduce the overall power consumption of the LP receiver.

The implemented LP DCO is shown in Figure 6.17. The architecture presented here uses the concept from [8] to lower the DCO consumption. The ring oscillator itself (transistors M_{7-11}) oscillates at one third of the

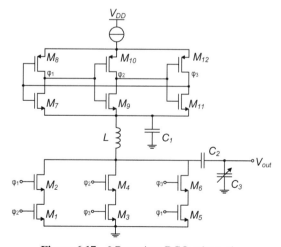

Figure 6.17 LP receiver DCO schematic.

desired frequency. The three phases ϕ_{1-3} are then combined using an edge combiner (transistors M_{1-6}) in order to multiply the output frequency by 3. The oscillator and the edge combiner are stacked on top of each other and reuse the same current. The two stages are separated by an LC filter that provides a stable voltage for the source of the ring oscillator NMOS transistors on one side, and high impedance for the combiner output on the other. The stable source voltage sets at approximately 0.35 V and does not change significantly with the oscillation frequency. The resonance frequency at the edge combiner output is set by the values of the inductor L and the capacitor C_3, and can be tuned by switching the capacitor bank C_3. The edge combiner acts at the same time as the LO buffer in the sense that a change of the load capacitance at its output does not affect oscillation frequency (in the first order approximation). No additional buffers are added before the mixer and frequency divider input. Oscillation frequency is controlled via the supply current of the ring oscillator. One downside of the chosen DCO implementation is the fact that output amplitude and oscillation frequency cannot be set independently. The oscillator bias current, is at the same time bias current of the frequency tripper and therefore also sets the output amplitude. As a consequence, the output amplitude is lower than in the case of the MU oscillator resulting in lower mixer conversion gain and hence in a higher noise figure.

Simulation results of the designed oscillator-tripper are shown in Figure 6.18. The oscillator covers a range from 3.5 GHz to 5 GHz, which for 6 control DCO bits corresponds to a resolution of around 25 MHz. At the same time power consumption varies from 57 μW to 88 μW. At 4 GHz the oscillator is expected to consume 65 μW from a 1 V supply. At this power consumption the DCO output amplitude is equal to 85 mV. As it can be seen in Figure 6.18, the amplitude is highly dependent on the resonance frequency of the output LC network, which was not the case in the MU DCO that provides an almost constant amplitude over the entire frequency range. The tuning capability is added, in order to compensate for process variations and precisely tune the resonance frequency and maximize the output amplitude at 4 GHz after production.

6.4.4 Demodulator

The demodulator of the MU receiver, shown in Figure 6.19, is similar to the previously implemented demodulator. The only difference is the load of the double balanced mixer, that here has a band-pass characteristic

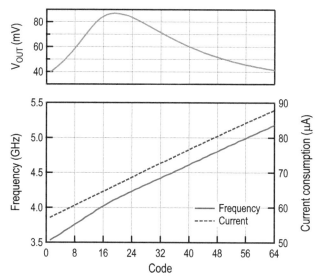

Figure 6.18 LP receiver DCO simulated frequency, current consumption and output voltage.

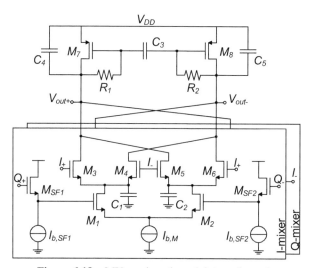

Figure 6.19 MU receiver demodulator schematic.

instead of a low-pass characteristic. In principle, this should improve the suppression of narrowband interferers. After the first FM demodulation the narrowband interferer should be located at very low frequencies, determined by the signal bandwidth, that should be filtered out. Capacitor $C_4(5)$ attenuates components at frequencies higher than 2.5 MHz, and doesn't play a role below 1 MHz. Disregarding this capacitor the mixer load impedance is given by

$$Z_{out} = \frac{1}{G_{m7}} \frac{1 + 2sR_1C_3}{1 + 2sC_3/G_{m7}}. \tag{6.3}$$

At low frequencies, the mixer output impedance will be low, and equal to $1/G_{m7}$, thus attenuating potential interferers. At frequencies above the high-pass cut-off frequency $\omega_H = G_{m7}/2C_3$, the impedance seen from the mixer increases to R_1, providing higher voltage gain.

The LP demodulator consists of the frequency discriminator and the envelope detector. The frequency discriminator is implemented as a high-pass filter and is a part of the IF amplifier. The envelope detector is shown in Figure 6.20. The circuit is essentially a double balanced mixer, where the input signal is mixed with itself. To provide the two different bias points for the two mixer inputs, two different source followers were used. Source followers $M_{SF1,3}$ use native NMOS transistors with a 0 threshold voltage and drive the first mixer input (transistors M_{3-6}). Lower bias for the second mixer input (transistors $M_{1,2}$) is provided by the low threshold voltage devices $M_{SF2,4}$. The bias currents are the same for all the source follower stages. For the load of the mixer, the same approach is used as for the MU demodulator, with the difference that the pass-band is set from 2 MHz to 2.5 MHz. This is done because the LP demodulator doubles the frequency of the sub-carrier signal. Since the transmit SC channel is centered at 1.05 MHz, the received sub-carrier signal is located at 2.1 MHz.

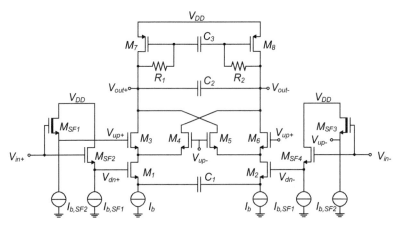

Figure 6.20   LP receiver demodulator schematic.

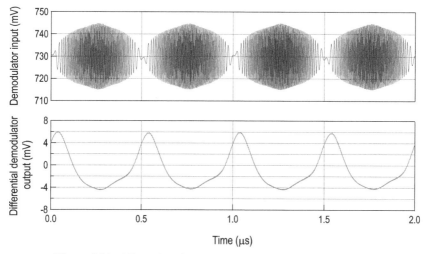

Figure 6.21 LP receiver demodulator input and output waveforms.

Simulated waveforms of the demodulator input and output signal are shown in Figure 6.21. The effect of the high-pass characteristic of the IF amplifier can be observed in the input signal as the low frequency components are highly attenuated. After self mixing, the high frequency components disappear, and only envelope is left at the output. The output pulses appear at twice the transmitted SC frequency. After passing through the low-pass filter that follows the demodulator, higher components are attenuated and signal resembles a sine wave at the comparator input.

6.4.5 N-Path Channel Filter

The output of the first FM demodulator is the FSK modulated sub-carrier signal. Depending on whether one or several transmitters transmit simultaneously, there may be one or more sub-carrier signals present at the demodulator output. The purpose of the channel filter is to amplify the desired channel and filter out all the interfering sub-channels. As explained previously, the implemented receiver targets 4 sub-channels, each 200 kHz wide, with 300 kHz separation between adjacent channels. The implemented receiver is targeting a maximum of 10 dB difference in power levels between the two input FM-UWB signals. This translates into 20 dB difference in power of sub-channels after the first FM demodulator. To provide sufficient SNIR (signal to noise and interference ratio) before the FSK demodulation, the filter should attenuate the interfering sub-channels by 40 dB, thus providing the desired signal 20 dB stronger than the interferer. This chosen constraint is somewhat more stringent than necessary in order to provide margin for slight performance degradation compared to simulations. The filter should therefore provide a 200 kHz pass-band, and attenuation of 40 dB at 250 kHz away from the center frequency. Furthermore, the filter must be tunable from 1 MHz to 2.2 MHz in order to cover the entire sub-carrier range.

N-path filters seem like an excellent candidate for the given specification as they are known for their wide tuning range and high quality factor. The principle of N-path filters has been known for a long time [9], but they gained significant popularity in recent years as an alternative solution for high-Q RF filters that does not require off-chip passive components. Typical N-path filter consists of N parallel branches, each of them containing a switch and a capacitor in series. By driving the switches with N non-overlapping clock phases with frequency f_{ck}, the structure acts as a band-pass filter with a center frequency f_{ck}. This can be seen as a low-pass to band-pass transformation of the filter made of the input resistance and the N aforementioned capacitors. The achievable Q-factor can be very high and is proportional to the RC constant, the number of phases used and the clock frequency [9–11]. There are some downsides to N-path filters, that are generally present in all sampling systems. First one is that the pass-band also appears at the integer multiples of the clock frequency. An improvement can be obtained by connecting the capacitors differentially, which removes all the even harmonics. The remaining harmonics still need to be filtered out by another filter following the N-path filter. The second downside is folding of signal and noise around frequencies that are multiples of Nf_{ck} (aliasing). This issue is typically solved by introducing an antialiasing filter that precedes the N-path filter, just like it is done with any sampling system (e.g. ADC). Fortunately, in this particular application noise is not an issue as this filter is close to the end of the receiving chain.

N-path filters have been used in many receivers, and have been proven to provide good linearity and interference rejection [12–15]. However, in all these implementations N-path filters act as a high-Q second order filter, with a very narrow pass-band (relative to the center frequency). In this application, a relatively flat pass-band characteristic is required, along with a linear phase, in order to avoid distortion of the sub-carrier signal. Ideally, this requires translation of a higher order equivalent low-pass filter to the desired center frequency. The described design has been demonstrated for RF frequencies in [16–18], here it is reused and adapted for low frequency and low power operation.

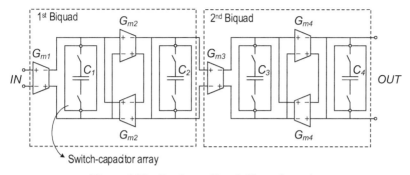

Figure 6.22 Band-pass N-path filter schematic.

The implemented N-path filter is presented in Figure 6.22. It is immediately clear that if the switches are removed, the shown filter becomes a standard low-pass G_m-C filter. In fact, it was shown in [18] that by adding switches, and scaling the capacitor values by the number of phases, such that $C_x = C_{BBx}/N$, the low-pass characteristic translates into an equivalent band-pass characteristic. This characteristic will be affected by the non-idealities such as switch resistance and parasitic capacitance. The parasitic capacitances will result in slight asymmetry around the filter center frequency. It is possible to compensate for the effect of the parasitic capacitances by adding feed-forward capacitors [18]. In this case, it is not necessary to add the compensation capacitors as the parasitics are relatively small compared to the actual filter capacitors due to narrow filter bandwidth and low frequency of operation. Switch resistance affects the quality factor of the filter and limits the attenuation in the stop-band. This is an important problem at RF, as the filter is typically driven by a 50 Ω source, since the maximum attenuation is limited to

$$A_{at,max} = \frac{2R_{sw}}{R_{in} + 2R_{sw}},\tag{6.4}$$

where R_{in} is the source resistance, and R_{sw} is the switch resistance, factor 2 is present in differential implementation, since two switches are used. In order to achieve the desired attenuation, switch resistance must be sufficiently low (much smaller than 50 Ω). This constraint will dictate the size of the switch and the driving requirements, and consequently power dissipation of the clock network. In this application, however, the filter is driven by an OTA with a high output resistance, on the order of 20 kΩ, which enables use of relatively small switches, and low power consumption.

The filter design procedure is done in two steps. First, a low-pass equivalent filter is designed, with a 100 kHz pass-band, and 40 dB attenuation at 250 kHz. In the second step, switches are added, and capacitors are scaled in order to obtain the desired 200 kHz pass-band characteristic around the center frequency. In other applications it might also be necessary to add feed-forward capacitors to compensate for the parasitics. For this design, it was determined that a 4th order, type 1 Chebychev transfer function satisfies the given specifications. Two biquadratic sections are used to implement the network with 4 poles p_{1-4}. The low-pass equivalent transfer function of each biquadratic section is given by

$$H(s) = \frac{H_0}{as^2 + bs + 1}, \tag{6.5}$$

$$H_0 = \frac{G_{m1}G_{m2}}{G_{m2}^2 + G_{o1}G_{o2}}, \quad a = \frac{C_1C_2}{G_{m2}^2 + G_{o1}G_{o2}},$$

$$b = \frac{C_1G_{o2} + C_2G_{o1}}{G_{m2}^2 + G_{o1}G_{o2}}, \tag{6.6}$$

where G_{o1} is the output conductance at the output of G_{m1}, and G_{o2} is the output conductance at the input of G_{m3}. The active filter can also provide some voltage gain. Assuming that $G_{m1}G_{m2}/G_{o1}G_{o2} \gg 1$, gain can be approximated as $A_v = G_{m1}/G_{m2}$. Coefficients a and b are determined by the two poles

$$a = \frac{1}{p_1p_2}, \quad b = \frac{p_1 + p_2}{p_1p_2}, \tag{6.7}$$

for the first biquadratic section, and in the same way, using p_3 and p_4, for the second biquadratic section. Once the coefficients a and b are set, parameters G_m, G_o and C must be chosen. The transconductances $G_{m1,2}$ are limited by the power consumption constraints that limit the bias current of each transconductor. Transconductors are implemented using a Krummenacher differential pair [19], as shown in Figure 6.23, in order to provide better linearity. Transistors of the differential pair are biased in weak inversion, with $\beta_1/\beta_3 = 2$, corresponding to the minimum ripple condition. With 4 μA per differential pair, the equivalent transconductance is set to 40 μS. Output resistance of each cell (R_1 from Figure 6.23), and capacitors $C_{1,2}$ are then determined to implement the desired transfer function. Finally, the chosen capacitance values are scaled by factor N, which corresponds to the number of phases.

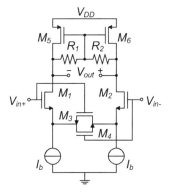

Figure 6.23 Transconductor of the N-path filter.

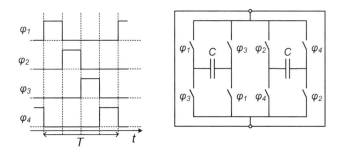

Figure 6.24 Non-overlapping clock phases used to drive switches and the differential switch-capacitor array.

In this design, four phases are used. The driving signals, and the switched-capacitor array are shown in Figure 6.24. Differentially connected capacitors cancel out the even harmonics. Using more phases would allow slightly better performance and less noise folding, but would increase the number of switches, and power consumption. Furthermore, four non-overlapping phases, can be generated using an input clock at four times the filter center frequency. In this case, it is convenient since the same clock can be used by the FSK demodulator. The circuit that generates different phases ϕ_{1-4} is shown in Figure 6.25. The three flip-flops divide the input clock by 4 and provide four equally spaced phases with a 50% duty cycle. The desired waveforms are then produced by the four AND gates.

Finally, the simulated filter characteristic is shown in Figure 6.26. In the simulation, 5 MHz input clock is used, resulting in center frequency of

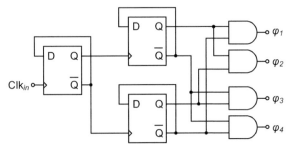

Figure 6.25   Non-overlapping clock generator.

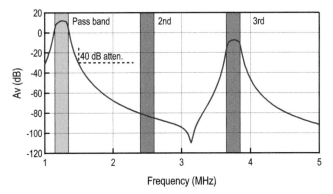

(a) Channel 1.

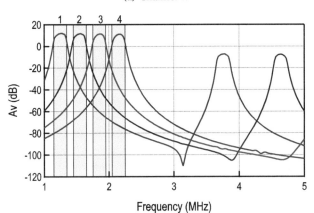

(b) All four channels.

Figure 6.26   Transfer function of the N-path filter.

1.25 MHz. As explained previously, the differential capacitors cancel out all the even harmonics, and so the first higher order harmonic appears at 3.75 MHz. This is off-course the idealized case, and the attenuation of second harmonic will be limited by the component matching. Slight asymmetry around the center frequency can be observed, as expected, but in this case it does not cause severe distortion of the FSK signal. The filter center frequency can easily be tuned by adjusting the input clock frequency. Furthermore, bias currents of the transconductors can also be adjusted to modify the gain, filter bandwidth and attenuation in the stop-band.

6.4.6 LF Amplifier and Comparator

The low frequency (LF) amplifier, placed before the comparator, filters out high frequency noise, and amplifies the signal to the level needed by the comparator. The same architecture is used in both MU and LP receivers, with the pass-band adjusted to the desired frequency range. In the MU receiver, this filter also attenuates the 3rd harmonic of the N-path filter transfer function. A cascade of two fully differential amplifiers is used. A single amplifier cell is shown in Figure 6.27, together with the small signal model of the half circuit. Each amplifier actually implements a 2nd order transfer function. The idea to use a negative resistance to implement the second order function was found in [20, 21]. This approach also allows a high quality factor, using a negative resistance that cancels out the real part of the output impedance, however,

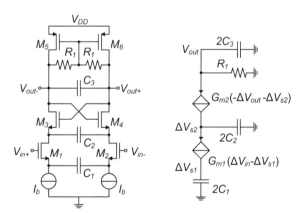

Figure 6.27 Schematic of the second order cell of the LP filter and half circuit small signal schematic.

one must be careful to maintain the circuit stable and avoid oscillations. The voltage gain is given by

$$A_v(s) = \frac{s\, 2C_1 G_{m1} G_{m2} R_1}{(G_{m1} + s2C_1)}$$

$$\times \frac{1}{(G_{m2} + s(2C_2 + 2R_1 G_{m2}(C_3 - C_2)) + s^2 4C_2 C_3 R_1)}. \quad (6.8)$$

The zero and pole created by C_1 provide the high-pass part of the characteristic that filters out flicker noise together with any low frequency components. The second factor in the denominator provides the 2nd order low-pass filtering. If the first real pole is sufficiently far from the two complex poles, gain in the pass band will be given by

$$A_{v,pb} \approx G_{m1} R_1. \quad (6.9)$$

For the two poles generated by this factor to be in the left half-plane, the coefficient with s must be positive. The circuit remains stable as long as $2C_2 + 2R_1 G_{m2}(C_3 - C_2) > 0$. This will be guaranteed if the capacitor C_3 is larger than the capacitor C_2. Otherwise, given the sufficiently high bias current, and consequently the transconductance G_{m2}, the circuit might start to oscillate.

The simulated frequency characteristics of the two LF amplifiers are given in Figure 6.28. The frequency band is selected based on the expected sub-carrier frequency. Since a larger band is needed for the MU receiver, in order to accommodate multiple SC channels, gain is slightly lower, around 25 dB in the pass-band, compared to approximately 32 dB in the LP receiver

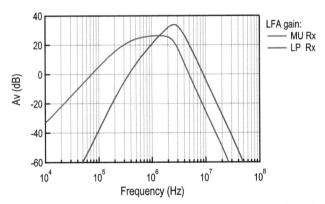

Figure 6.28   Simulated frequency characteristic of the MU and LP receiver LFA.

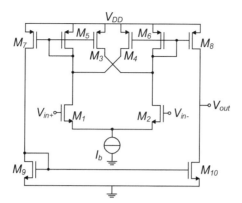

Figure 6.29   Comparator schematic.

path. In this case, linearity is not a concern, since only a single FSK signal is expected at the input (additional FSK signals should be removed by the preceding channel filter).

Comparator following the LF amplifier acts as a limiter and provides a rail-to-rail output signal that is needed for the digital FSK demodulator. The schematic of the comparator is shown in Figure 6.29. It is designed to provide full swing for an input sine signal with a minimum differential amplitude of 20 mV at up to 5 MHz (which is above the needed range). The core of the comparator are transistors M_{1-6}. In order to provide better performance and avoid glitches due to noise, a small hysteresis is introduced in the comparator characteristic, on the order of 10 mV. This is done using the positive feedback transistors M_3 and M_4. Assuming the transistors are operating in weak inversion, the difference between the two threshold voltages is given by

$$\Delta V_{TH} = 2nU_T \ln k, \qquad (6.10)$$

where factor k is defined as $k = \beta_3/\beta_5 = \beta_4/\beta_6$. Transistors M_{7-10} are added to provide a full swing output signal compatible with CMOS logic. The comparator consumes between 5 μA and 10 μA, depending on the bias current setting. Since the expected input voltage amplitude is supposed to be larger than 20 mV, offset constraints are easily achievable and no calibration is necessary.

6.4.7 FSK Demodulator and Clock Recovery

The last block in the system is the FSK demodulator, that is implemented together with the clock recovery circuit. The demodulator implemented here must be able to demodulate an FSK signal with the modulation index of 1, or equivalently 50 kHz frequency deviation at a data rate of 100 kb/s. The FSK demodulator reported in [3], is very simple, and consumes a small amount of power, but requires a large frequency deviation (250 kHz deviation was used). In this case, in order to support multiple sub-carrier channels, frequency deviation is limited to 50 kHz, and a different approach is needed.

The proposed demodulator, shown in Figure 6.30 is a digital version of the delay line demodulator. The input signal is first sampled using a clock whose frequency is four times higher than the FSK signal center frequency. The same clock is used for the N-path channel filter. By adjusting the reference clock frequency, the corresponding sub-channel is selected. The sampled signal is then demodulated using a delay line and a "mixer". Sampling the signal allows to implement the delay line as a chain of flip-flops controlled by the same reference clock. Delay can be configured easily by controlling the number of flip-fops in the signal path, which is simply achieved by configuring the multiplexers. This allows a more elegant control compared to analog solutions such as RC delay networks. An XOR gate plays the role equivalent to the mixer in the analog demodulator implementation. Depending on the delay, and whether the input frequency is higher or lower than the reference clock frequency, the output of the XOR gate will be '1' or '0'. Simulated signal at the FSK demodulator output is shown in Figure 6.31(a). Since the output signal is not perfectly clean (even without the presence of noise), it cannot be simply sampled, instead it is first filtered using a windowed accumulator. In each clock cycle, the accumulator output is either incremented or decremented depending on the XOR output. The accumulator output is then used to make a decision for the output bit and to recover the symbol clock.

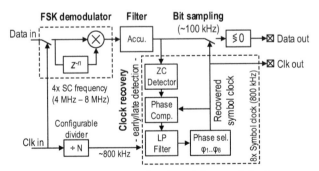

Figure 6.30 Block diagram of the FSK demodulator and clock recovery circuit.

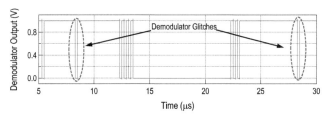

(a) FSK demodulator output signal.

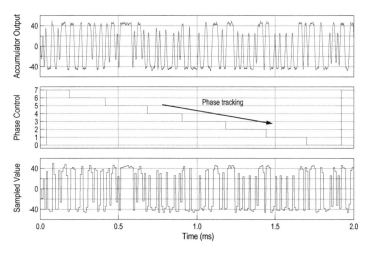

(b) Internal signals of the clock recovery circuit.

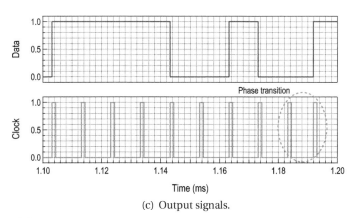

(c) Output signals.

Figure 6.31 Simulated signals of the FSK demodulator and clock recovery circuit.

A clock recovery circuit is a necessary block in any receiver, and has a particularly important role here. As previously discussed, one of the good properties of FM-UWB is the inherent robustness to frequency offsets. Mismatch in carrier frequency of several megahertz, or even tens of megahertz will not cause a significant performance penalty. However, for the whole receiver chain to work properly, the baseband must also be able to tolerate a certain frequency offset. This is accomplished via a dedicated circuit that tracks the transmit symbol clock and adjusts the clock frequency on the receiver side. The amount of frequency offset that must be tolerated depends on the implementation of the reference oscillator. One of the aims of this work is to demonstrate the feasibility of a fully integrated transceiver, which would include the reference oscillator. This also means removing the external crystal reference and minimizing the number of off-chip components. Unfortunately, the integrated RC oscillators cannot achieve the performance of a crystal oscillator, and the precision of the reference frequency will be much worse, in the order of thousands of ppm instead of tens of ppm. By using the FM-UWB, this relatively large frequency variation can be allowed, assuming that the clock recovery circuit can compensate the frequency offset between the transmitter and the receiver. Recently, integrated RC oscillators achieved precision that is in the order of ± 2500 ppm across the designated temperature range [22–25], which is a range that can be easily covered by the clock recovery circuit shown here.

The implemented clock recovery is based on a simple early/late zero crossing detection. The clock used for this circuit is derived from the reference clock, with the average frequency 8 times higher than the symbol rate, that is 800 kHz. This clock is then used to generate 8 different phases of the symbol clock, one of which is used to sample the accumulator output at a correct time instance. The clock recovery circuit determines which phase is used and works in the following way. First a zero crossing is detected from the accumulator output. In order to avoid false crossings due to noise, the circuit needs to detect a sufficient difference in levels between several consecutive samples. The phase comparator then determines whether the current zero crossing is early or late with respect to the currently selected clock phase. Depending on the number of clock cycles between the zero crossing and the reference, the corresponding value will be added to or subtracted from the register value of the LP filter (here the LP filter is simply implemented as an accumulator). Once the register value increases or decreases past a defined point, an up or down phase shift occurs. Depending on the frequency offset between the transmitter and the receiver, phase shifts will occur more or less often. As the phase changes, so does the average frequency of the receiver symbol clock. As long as the circuit is able to track symbols, this average frequency should correspond to the transmitter symbol clock frequency. The speed of the control loop can be controlled through the LP filter coefficient, that directly determines the filter bandwidth. Increasing the coefficient allows the loop to track larger difference in frequencies, but also makes it more prone to errors due to noise.

An example of the clock recovery circuit operation is given in Figure 6.31(b). In this case, the transmitter reference frequency is 2000 ppm faster compared to the reference on the receiver side. One can notice that the phase control signal constantly decreases (until 0 at which point it goes back to 7), which results in the average frequency of the symbol clock below the reference frequency. An example of the phase shift is shown in Figure 6.31(b). In that particular time instance the instantaneous frequency of the symbol clock frequency drops to 7/8 of the reference frequency during one cycle. The maximum theoretical frequency offset that can be tracked, assuming a phase shift occurs in every cycle, is ±1/8 of the reference frequency.

6.4.8 SAR FLL Calibration

Ring oscillators used to generate the LO signal in the receiver and the FM-UWB signal in the transmitter consume a small amount of power, but are sensitive to process, voltage and temperature variations. For that reason they need to be calibrated periodically to maintain frequency offset within certain limits. It was shown in the previous chapter that a relative frequency offset of ±50 MHz between the receiver and transmitter causes only a minor performance degradation, and beyond that limit sensitivity decreases rapidly. Depending on the rate of environmental changes, the oscillators will need to be calibrated once every few hours, or potentially days.

Calibration is performed using an on-chip SAR FLL, shown in Figure 6.32. Configuration of each oscillator is controlled using one of the two registers, one set manually by the SPI, and the other set by the calibration loop. The number of cycles for calibration is equal to the number of register bits used to set the DCO frequency. In each cycle one bit is set. Once the bit is set, the oscillator frequency is measured and compared to a reference value, if it is higher, the bit is set back to '0', otherwise it remains '1'. The DCO frequency is measured using two counters. The first counter counts the number of reference clock cycles up to value $N_{f,ref}$ that determines the duration of the measurement interval as $T_{ref}N_{f,ref}$. During that interval, the second counter counts the number of cycles of the frequency divider output N_{cnt}. This value is then compared to the reference value N_{ref} in order to determine

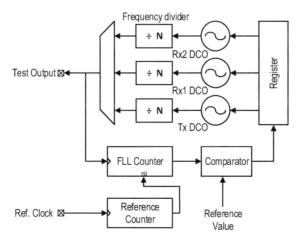

Figure 6.32 SAR FLL block diagram.

whether the corresponding bit should be '0' or '1'. Finally, assuming infinite resolution (number of bits or equivalently cycles), the frequency of the DCO after calibration will be given by

$$f_{DCO} = N_{div}f_{div} = N_{div}\frac{N_{ref}}{N_{f,ref}}f_{ref},\qquad(6.11)$$

where N_{div} is the frequency divider ratio (providing values between 128 and 1024) and f_{div} is the frequency at the divider output. In reality, the DCO configuration will produce the highest frequency that is still below the frequency given by Equation 6.11. The frequency resolution of the FLL depends on the duration of the measurement interval and the division ratio $\Delta f = f_{ref}N_{div}/N_{f,ref}$. The only requirement here is to maintain the FLL resolution below the DCO resolution, that is in the order of 25 MHz.

An example of a measured calibration cycle is presented in Figure 6.33. The measurement is done using the frequency divider output signal at the test port. For the shown measurement, reference clock frequency is set to 4 MHz, and $N_{div} = 1024$, $N_{f,ref} = 1024$ and $N_{ref} = 1000$. After 6 steps of calibrations, 6 DCO control bits are set. The resulting output signal of the frequency divider is at 3.89 MHz, corresponding to the DCO frequency of 3.98 GHz, which is close enough to the ideal carrier frequency.

6.4.9 Clock Reference

The implemented FM-UWB transceiver does not contain an oscillator that would provide a frequency reference for other circuits (LO calibration,

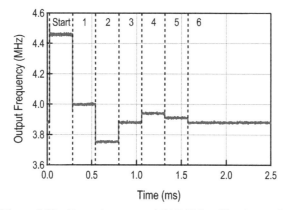

Figure 6.33 Example measured SAR FLL calibration cycle.

N-path filter, FSK demodulator etc.). Instead, the clock signal is generated externally using a signal generator, allowing to adapt to different modes of operation and test the functionality of the transceiver. Since one of the goals of this work is to show that an FM-UWB transceiver can be implemented without a precise, off-chip quartz resonator, one solution for derivation of all the necessary clock signals will be described here.

A simple RC oscillator could be used as a clock reference, similar to one of the solutions found in [22–25]. State of the art precision of 2500 ppm across the temperature range of interest is sufficient for the FM-UWB transceiver developed here. The FSK demodulator and the clock recovery circuit were designed to compensate for the potential frequency offset. At RF this range translates into ± 10 MHz offset around the carrier frequency of 4 GHz, which is well bellow the targeted range of ± 50 MHz. The RC would simply provide the fixed reference clock signal, different frequencies needed for different circuits would be generated using a simple FLL or a DLL (delay locked loop). The block diagram of the conceptual solution with all the related sub-blocks is shown in Figure 6.34.

For the SC-DDS, a clock frequency of at least 40 MHz is needed to provide a relatively good triangular signal. Higher frequencies should provide a better triangular waveform, but would also result in increased power consumption of this block. At the same time the FLL must be able to provide frequencies from 4 MHz to 8 MHz, as a reference for the N-path filter and the FSK demodulator. Luckily, the same frequency is used by both circuits. For the N-path filter, the clock is divided by 4 in order to generate the four non-overlapping phases at the sub-carrier center frequency. The symbol clock

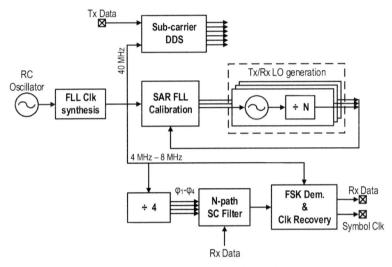

Figure 6.34   Principle of clock generation and distribution.

of approximately 100 kHz is derived from the input clock using a simple counter, and the clock recovery circuit assures that it tracks the symbol rate of the received signal. No particular constraints in terms of input frequency exist for the FLL calibration loop, the two configurable reference values (see Figure 6.32) can always be set such that the DCO frequency is properly calibrated. The proposed clocking scheme can therefore be used to provide a reference to all the circuits in the system, demonstrating one way to implement a fully integrated FM-UWB transceiver.

6.5 Measurement Results

The proposed transceiver was integrated in a standard 65 nm bulk CMOS technology. The SEM die photograph is shown in Figure 6.35. The die size is 2.25 mm by 2.25 mm, and roughly one third of it is the active area of the transceiver (including the decoupling capacitors). The remaining area is used for test circuits and decoupling capacitors. The transceiver layout is dominated by the inductors needed to provide input and output matching. Large inductor area makes routing more difficult and requires longer paths, that consequently add more parasitics at the transceiver IO. It should be noted here that standard TSMC inductors were used for the design. These inductors use only a single metal layer (the low resistance ultra-thick metal), and hence occupy a large area. The layout could be made more compact using smaller,

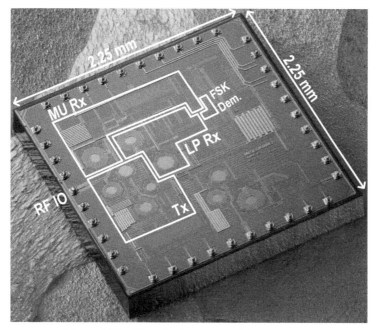

Figure 6.35 SEM die photograph of the transceiver.

more area efficient custom inductors, that would exploit additional available metal layers. The RF IO pad, used by the receivers and the transmitter, is marked in the figure. It is placed between the two ground pads, that are connected to the coplanar waveguide implemented on the PCB. Digital input and output pads, used for test and debug signals, clocks and input and output bits, are located on the side of the chip opposite to the RF IO pad and other sensitive analog signals in order to minimize coupling between strong digital signals and sensitive analog signals. Gold bumps, used for flip-chip bonding of the IC to the PCB, can also be seen in Figure 6.35.

6.5.1 Transmitter Measurements

The first block of the transmitter is the sub-carrier DDS. The static configuration that controls the two sub-carrier frequencies is loaded via the SPI, and the dynamic behavior of the circuit is controlled using the two inputs, the clock and the data input. The SC-DDS controls two current steering DACS, one that drives the DCO, and the other one that drives the test buffer. The output signal from the test buffer is presented in Figure 6.36(a). The shown waveform corresponds to 2.1 MHz sub-carrier signal and 500 MHz wide

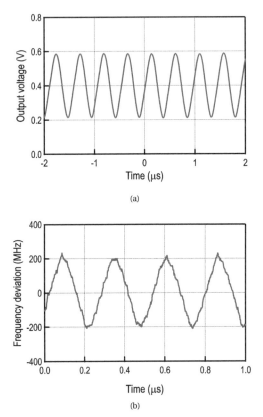

Figure 6.36   Measured sub-carrier DAC output (a) and measured frequency deviation of the transmitted signal (b).

FM-UWB signal. Deviation from the triangular waveform is a result of the limited bandwidth of the buffer. Sub-carrier waveform can also be obtained by directly measuring instantaneous frequency of the transmitted signal. The frequency measurement is done using the Keysight VSA application with the 8 GHz, MSO oscilloscope. The resulting waveform is depicted in Figure 6.36(b). Due to large input bandwidth, the resulting signal is relatively noisy, but still shows some of the properties of the generated SC signal. In this case SC frequency is set to 4 MHz, which is in fact above the targeted SC frequency range. It can be seen in the figure that the output waveform deviates from a triangular close to the peaks. As a result the bandwidth of the output FM-UWB signal will be slightly lower than expected. This is a result of the low oversampling ratio (ratio between the clock frequency and the sub-carrier

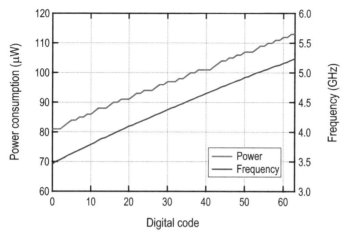

Figure 6.37 Frequency and power consumption of the transmit DCO.

frequency f_{clk}/f_{SC}). In the presented example, the oversampling ratio is 10, for a 40 MHz input clock. According to [2], a relatively good result is obtained for an oversampling ratio of around 20, which will be the case for the desired SC band (1.2 MHz-2.3 MHz).

The measured DCO frequency and power consumption are presented in Figure 6.37. The DCO frequency can be varied from 3.5 GHz to 5.2 GHz and for this range the current consumption varies from 80 μA to 113 μA. This frequency tuning range is achieved using the static DAC that only sets the upper FM-UWB frequency. Dynamic modulation DAC then sinks the current from the static DAC to modulate the carrier frequency. The reported measurement is the consumption of the DCO alone, without the buffer consumption. For the nominal setting, the buffer consumes an additional 71 μA of current. By changing the buffer bias current, the DCO output amplitude can be adjusted, and the consumption can be varied from 48 μA to 94 μA. The input signal amplitude allows to control the output power and current consumption of the PA and PPA, and optimize the transmitter efficiency.

Due to the large bandwidth, FM-UWB signal is inherently robust against carrier offsets. The same property allows it to tolerate relatively high levels of phase noise. It was shown in [3] that phase noise as high as -80 dBc/Hz, at 10 MHz away from the carrier, causes no significant performance degradation in terms of BER. This constraint is quite loose and permits the use of low quality ring oscillators for signal generation. For comparison, consider the Bluetooth standard that imposes a constraint of -102 dBc/Hz

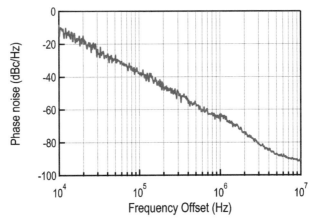

Figure 6.38   Phase noise of the transmit DCO at 4 GHz.

at 2.5 MHz [26], and consequently requires higher power consumption for carrier synthesis. The measured transmit oscillator phase noise, for the oscillation frequency of 4 GHz, is shown in Figure 6.38. At 10 MHz away from the carrier, phase noise level is −98 dBc/Hz, which is considerably lower than the FM-UWB constraint. The phase noise was measured using the signal at the output of the frequency divider. A factor of 20 log 1024 was added to the measured phase noise to account for the division ratio of 1024. Noise coming from the dividers will add to the total phase noise at the output; however, due to the already large phase noise of the ring oscillator it should not have a significant impact on the measured value.

The power amplifier and the output matching network were designed to provide good performance over the entire 500 MHz range. The idea is to achieve high average efficiency of the transmitter during wideband signal transmission, and not at a single frequency, which is usually the approach in narrowband systems. Static frequency characteristic of the PA and PPA stack is summarized in Figure 6.39. The measurement is conducted using the DCO as the input signal source, since an external signal cannot be used. For this measurement the DCO is configured to produce a carrier signal at a single frequency. Due to high phase noise and unstable frequency of the ring oscillator, the result is likely worse than it would have been if a clean carrier signal were used.

The measured output power of the transmitter is shown in Figure 6.39(a). Compared to simulations, the level is approximately 2 dB lower, with a somewhat smaller bandwidth. The discrepancy between the simulation and

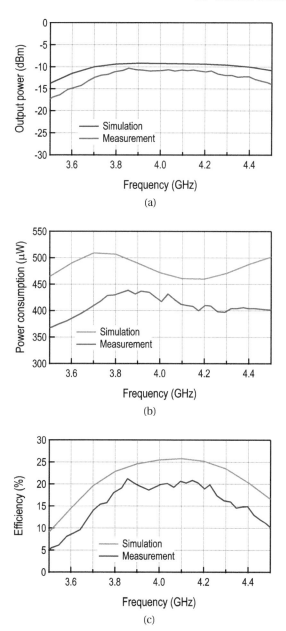

Figure 6.39 Measured power amplifier output power (a), consumption (b) and efficiency (c) including the preamplifier.

measurement is likely caused by the effects that were not taken into account in the simulation. A simplified model of the output pad and the interface toward the PCB were used, more accurate results would have been obtained by using a full 3D electromagnetic simulation. Nevertheless, the error remains within the expected limits of a few decibels. Fortunately, the impedance of the matching network also affects the power consumption. The fact that it slightly differs from the simulated values also results in decreased measured power consumption compared to the simulated one, as shown in Figure 6.39(b). Lower output power, combined with lower power consumption, finally result in efficiency around 6% lower than expected in simulations. As shown in Figure 6.39(c) peak measured efficiency of the output stage (not including the DCO and the buffers) equals 21.3%. In the largest part of the band efficiency stays around 20%, and slowly drops close to the edges of the band, resulting in an average efficiency of around 18%. The measured output power and efficiency of the implemented power amplifier are in line with the state of the art and exhibit similar performance as other implementations targeting low output power levels.

The spectrum of the transmitted signal is presented in Figure 6.40. The shown spectrum is below the limit defined by the FCC spectral mask. Outdoor spectral mask is shown in this case, the only difference between the outdoor and the indoor mask being the attenuation outside the defined UWB band, which is more stringent for the outdoor mask. The transmit signal also satisfies the defined emission level between 0.96 GHz and 1.61 GHz, where maximum level set to -75 dBm/MHz (not shown in Figure 6.40).

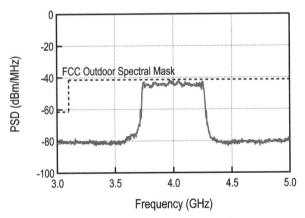

Figure 6.40   Transmitted FM-UWB signal spectrum.

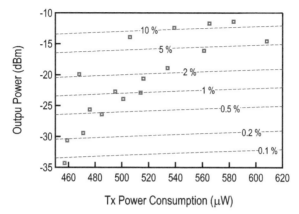

Figure 6.41   Transmit power vs. transmitter power consumption.

Figure 6.41 shows different output power levels and average transmitter power consumption, achievable using different transmitter configurations. Constant efficiency lines are provided to show the achievable efficiency at a given output power. In this case, the whole transmitter is taken into account including the DCO, buffers and SC-DDS (not just the PA and PPA). The shown output power levels are measured as integrated power across the transmit band during a single sub-carrier transmission. The output power level can be varied from -11.3 dBm down to -35 dBm, in steps that are smaller than 3 dB. This is done via the control of the DCO buffer strength (or equivalently PPA input signal amplitude), the PPA bias current and the matching network. The control of the output power can then be implemented using a look-up table in software. The ability to adjust the output power is useful for multi-user communication. For example, if the two transmitting nodes are at different distances from the receiving node, the power levels of the two signals will be different. In order to equalize power levels at the receiver, the closer transmitter can adjust its output power to avoid desensitizing the receiver.

Measured input S_{11} parameter is shown in Figure 6.42. The measurement is done using the die bonded to the PCB, with a 1 cm long coplanar waveguide between the RF IO pad and the horizontal SMA connector. The lower values compared to the simulations are likely related to the PCB, since the connector and the transmission line were not taken into account. On the receiver side, the additional losses coming from the PCB actually improve the reflection coefficient, that is below -10 dB from 3.6 GHz to 4.75 GHz.

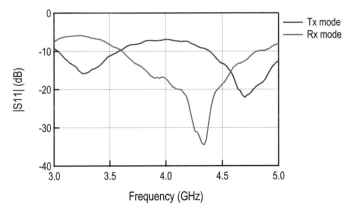

Figure 6.42   Measured S_{11} parameter in transmit and receive mode.

Table 6.1   Transmitter power consumption breakdown

Block	Current Cons. (μA)	Relative Cons. (%)
PA+PPA	402	69.9
DCO	91	15.8
DCO Buffer	71	12.4
SC DDS	11	1.9
Total	575	100

The power consumption breakdown of the transmitter is given in Table 6.1, for the transmitted power of -11.4 dBm. Almost 70% of the power is consumed by the output stage (PPA and PA). In order to improve the efficiency of the transmitter, this part should be carefully optimized in the future. The SC DDS consumption is practically negligible compared to other blocks, and so its implementation will have little impact on the overall performance. The DCO together with the buffers consume slightly less than 30%, thanks to the fact that there is no continuous time PLL or FLL controlling the output frequency.

6.5.2 Receiver Measurements

The two receivers are mainly characterized in terms of BER and sensitivity under different conditions. Unfortunately, there is no way to access different internal points of the receiver and characterize each block separately. Addition of buffers that would allow this would result in increased capacitance in the corresponding nodes, which would consequently increase the receiver

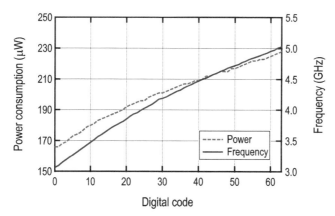

Figure 6.43 Measured frequency and power consumption of the MU Rx DCO.

power consumption. The buffers and test outputs are therefore added only at lower frequencies where such capacitive load causes no significant problems.

The DCOs of the two receivers are the most important blocks in the chain, since without downconversion it would be impossible to perform demodulation. Just like with the transmitter DCO, the frequency dividers are added to provide information about the DCO frequency and to close the FLL calibration loop. Unlike the transmitter DCO, linearity of the frequency characteristic is not needed on the receiver side. What is important is its monotonicity, that ensures proper operation of the SAR FLL loop. The frequency and power consumption of each DCO are measured using the frequency dividers and a digital output buffer. The result for the MU receiver is shown in Figure 6.43. The provided result also includes the four DCO buffers for quadrature LO signals. The oscillator frequency ranges from 3 GHz to 5 GHz, while power consumption changes from 166 μW to 228 μW. The resulting frequency resolution changes from 35 MHz to 20 MHz as the oscillation frequency increases. The same measurement for the LP receiver is given in Figure 6.44. Again, measured power consumption includes the buffer, or in this case the frequency trippler, since it reuses the current from the oscillator. In this case, the output frequency takes values from 3.6 GHz to 5.25 GHz, while it consumes between 51 μW and 77 μW. Due to nonlinear behavior of the oscillator, the frequency step reduces from around 30 MHz at the lower end, to 20 MHz at the high end.

Another block that can be measured standalone is the N-path filter. Figure 6.45 shows the voltage gain characteristic of the filter. Used input clock frequency is 5 MHz, which results in the filter center frequency of

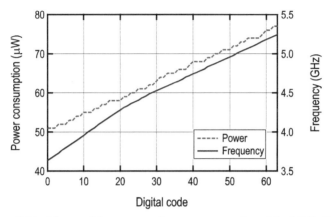

Figure 6.44   Measured frequency and power consumption of the LP Rx DCO.

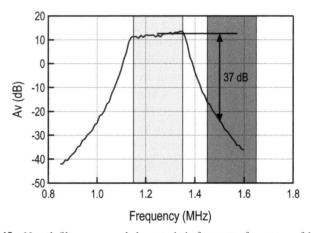

Figure 6.45   N-path filter measured characteristic for center frequency of 1.25 MHz.

1.25 MHz. For the given configuration, the filter provides 200 kHz of band-width, and attenuation of 37 dB at the frequency of the adjacent sub-carrier, which is 250 kHz away from the filter center frequency. As expected, the characteristic in the pass-band is not entirely flat. A small inclination appears as a result of parasitic capacitances in the layout, however, for this particular case the performance should not be affected. The purpose of the N-path filter is to remove the interfering sub-carrier channels. As a demonstration, the spectrum before and after the filter is shown in Figure 6.46. Signal spectrum after the wideband FM demodulator is shown in top part of the figure. Four FM-UWB signals of equal power are present at the receiver input. After demodulation,

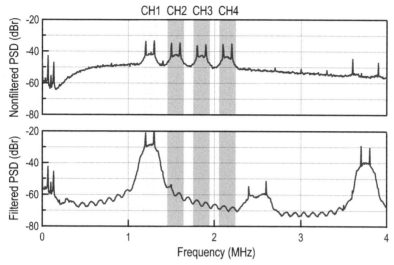

Figure 6.46 Demodulated signal spectrum before and after N-path filter.

four SC channels can be distinguished in the spectrum before filtering. After passing through the filter only channel 1 remains (bottom). Component at 3.75 MHz is a consequence of the sampled nature of the system. With the clock frequency of 1.25 MHz and 4 phases, the equivalent sample rate will be 5 MHz. Attenuated copy of the signal spectrum therefore appears at the frequency $(N - 1) f_{clk}$, where n is a number of phases, which is 3.75 MHz. Another visible component is the second harmonic of the SC channel at 2.5 MHz. This component is the combination of the output buffer non-linearity, and mismatch of the N-path filter. Ideally, second harmonic should be completely suppressed by the differential architecture of the N-path filter, however, in practice the amount of attenuation will be limited by matching.

Power consumption breakdown for the MU receiver is given in Table 6.2. As expected the highest amount of power is consumed by the high frequency blocks, the active mixer and the DCO, that together consume around 60% of the entire receiver consumption. The dominant consumer still remains the DCO, with buffers included, that consumes 194 μW. Among the low frequency blocks, notable amount of power is used for the N-path filter that provides sharp band-pass filtering. An overhead necessary to provide the multi-user capability. Finally, the total consumption of the MU receiver is 550 μW. This is more than the receiver presented in the previous section, mainly due to the fact that the baseband processing is now placed on chip.

Table 6.2   MU receiver power consumption breakdown

Block	Current Cons. (μA)	Relative Cons. (%)
LNA/Mixer	153	27.8
DCO	79	14.4
DCO Buffer	115	20.9
IFA	113	20.5
FM Demodulator	13	2.4
Channel filter	43	7.8
LFA	6	1.1
FSK Demodulator	8	1.5
Bias	20	3.6
Total	550	100

Table 6.3   LP receiver power consumption breakdown

Block	Current Cons. (μA)	Relative Cons. (%)
LNA/Mixer	103	38.6
DCO	53	19.9
IFA	46	17.2
FM Demodulator	23	8.6
LFA	11	4.1
FSK Demodulator	14	5.2
Bias	17	6.4
Total	267	100

Power consumption breakdown for the LP receiver is given in Table 6.3. The strategy with power reduction is to reduce power of some of the main consumers from the MU receiver. First, the DCO, that is now single-ended, consumes slightly more than one quarter of the MU DCO consumption, that is 53 μW. One downside of the LP oscillator is the lower output amplitude that will affect the sensitivity. The active mixer, consumes a comparable amount of power, since the architecture is the same, with the only difference that a single differential signal is used at the output (there are no I and Q branches). A second significant power saving is coming from the IF amplifier. Since there is no need for two branches, in the LP receiver the consumption is practically halved compared to the MU receiver case. The IFA consumes 46 μW instead of 113 μW in the MU receiver. The total power consumption adds up to 267 μW for the whole LP receiver chain.

The measurement setup used for the BER measurements is shown in Figure 6.47. The transmit bits and input vectors for signal generators are

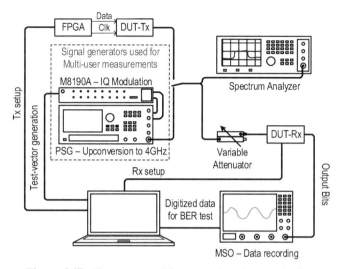

Figure 6.47 Test setup used for transceiver characterization.

created using Matlab. A configured FPGA then provides clock and data inputs for the transmitter. The spectrum analyzer is used to verify that the proper signal is generated by the transmitter and to measure the signal level at the receiver input. Multi-user measurements are still done using the signal generators. This simplifies the setup and allows to precisely control the relative power levels of different FM-UWB signals. The M8190A AWG provides the quadrature baseband signals, which are then up-converted by the PSG. The MSO oscilloscope is used to capture data at the receiver output.

All the BER curves are measured in two ways, first using the offline data post processing, and then using the on-chip FSK demodulator. This allows to measure the loss of the on-chip FSK demodulator compared to the ideal demodulator implemented in software. For the offline BER measurement, the MSO acts as a 20 MS/s, 10 bit ADC that captures analog data after the wideband FM demodulator. This data is then processed using software. First, the desired channel is filtered using an FIR band-pass filter to remove any undesired adjacent sub-channels. Then it is passed through a correlator, that is an optimal detector for the orthogonal FSK modulation used. The output bits are then compared to the generated test vector. In the second measurement, that uses the on-chip FSK demodulator, digital outputs are available, logic analyzer is used to sample data on a rising edge of the recovered clock. It is worth mentioning that the second measurement takes significantly less time (and memory) since the number of samples is significantly lower and corresponds to the number of bits. In the first case 200 samples are captured per received bit in order to perform the demodulation in software. This corresponds to a sampling frequency roughly 10 times higher than the sub-carrier frequency.

The BER curves for the MU receiver and the single user case are presented in Figure 6.48. The curves for both external (software) and internal demodulator are shown. In the case of the external demodulator the sensitivity is -69 dBm. The result is similar to the previous receiver implementation, with some losses due to a more complex input matching network. The internal demodulator adds 1 dB loss compared to the ideal software demodulator, resulting in a receiver sensitivity of -68 dBm. Accounting for approximately 2 dB loss due to a non-ideal LO, and 1 dB loss due to a non-ideal FSK demodulator, this is roughly 7 dB worse than the theoretical result.

The BER curves for the LP receiver in the single user case are presented in Figure 6.49. Just like in the above case, both curves, for the external and internal demodulator, are shown. In the case of the external demodulator the sensitivity is -58 dBm. Accounting for the theoretical 6 dB difference in sensitivities between the two receivers, and higher noise figure of the LP receiver, the 11 dB degradation in sensitivity is expected. Compared to the theoretical sensitivity, same 7 dB degradation is observed as in the case of the MU receiver. This is the price to pay for lower power consumption of the LP receiver. Again, the simple, low power internal demodulator adds some loss,

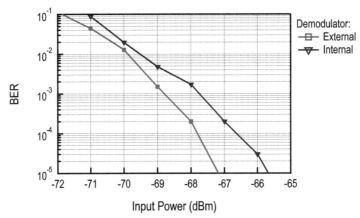

Figure 6.48   Single user BER of the MU Rx with internal and external demodulator at 100 kb/s.

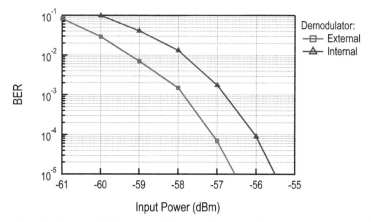

Figure 6.49   Single user BER of the LP Rx with internal and external demodulator at 100 kb/s.

resulting in −57 dBm sensitivity of the LP receiver. Although sensitivity is relatively low compared to other FM-UWB receivers, it is still enough for short range communication in a WBAN, and the implemented receiver consumes the lowest amount of power among all the implementations reported in literature.

A photo of the measurement setup is shown in Figure 6.50. Two boards can be seen in the figure, one is used as a transmitter and the other one as a receiver. The signal level on the receiver side is controlled using a configurable attenuator. Spectrum analyzer, showing the FM-UWB signal spectrum, is used to verify the proper operation of the transmitter and can be seen in the right part of the figure. In this case, the oscilloscope is used to compare the transmitted and the received bits. The recording from the screen is shown clearly on the graph. Aside from the input and output data, the graph also shows the recovered clock used for sampling the output data. A delay of approximately 20 μs can be seen between the transmit and receive bit stream. In the shown example there are no errors present at the output.

The sensitivity of the receiver degrades in the presence of interferers. The behavior of the receiver is evaluated in the presence of a narrowband interferer inside and outside of the used FM-UWB band. The interferer is generated using a separate signal generator and the outputs are summed together using a power combiner. For the in-band interferer, a frequency of 4.1 GHz was chosen as the worst case. Placing the interferer close to the signal center frequency would attenuate it due to the high-pass characteristic of the IF amplifier, and placing it closer to edge of the band would again

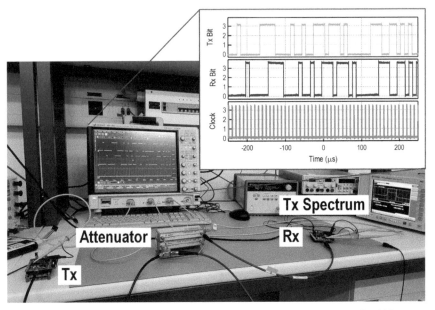

Figure 6.50   Measurement setup and comparison of transmit and received bits.

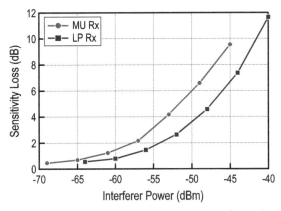

Figure 6.51   Sensitivity degradation due to the presence of an in-band interferer.

result in slightly lower IF gain due to the IFA low-pass behavior. The same frequency was used for both MU and LP receiver. The sensitivity degradation with the increase of interferer power is shown in Figure 6.51. Assuming 3 dB sensitivity degradation is acceptable, the MU and LP receiver can tolerate up to -55 dBm and -52 dBm strong interferers respectively.

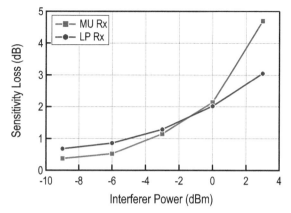

Figure 6.52 Sensitivity degradation due to the presence of an out of band interferer at 2.4 GHz.

For the out of band interferer case, the frequency of 2.4 GHz was chosen. The reason is that the 2.4 GHz ISM band (industrial, scientific and medical) is commonly used by different short range BAN devices. The sensitivity degradation in the presence of an out of band interferer is shown in Figure 6.52. Assuming again 3 dB sensitivity loss is acceptable, the MU receiver can tolerate 1 dBm interferer and the LP receiver can tolerate a 3 dBm interferer at 2.4 GHz. This is much higher than any other implementation and is due to the sharp filtering characteristic of the IF amplifier, that acts as a 6th order low pass filter. In most other implementations the interferer is only attenuated by the 2nd order LNA input matching network. This means that the proposed FM-UWB receivers can operate reliably next to any other device using the ISM band (such as a BLE radio for example).

As discussed previously, FM-UWB is inherently robust against frequency offsets thanks to its large bandwidth. This is clear when it comes to tolerance to LO signal offset, however the property also applies to other parts of the system. After the wideband FM demodulation, the sub-carrier channels are located between 1.2 MHz and 2.3 MHz. An offset of 1000 ppm translates into a maximum sub-carrier offset of 2.3 kHz, which would still not prevent correct FSK demodulation. For comparison, assume the same 1000 ppm offset is present in a Bluetooth transmitter that operates in the 2.4 GHz band. This would translate into an offset of 2.4 MHz, which is larger than a Bluetooth channel bandwidth. The example illustrates why an FM-UWB system has an advantage compared to typical narrowband systems. In fact, the FM-UWB receiver can tolerate a frequency offset much

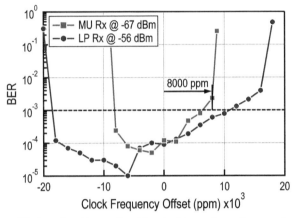

Figure 6.53   BER for a fixed input signal level with varying reference clock frequency.

larger than 1000 ppm. Figure 6.53 shows the BER of the two receivers as a function of the receiver reference clock offset. The curves were measured at 1 dB below the sensitivity level, that is −67 dBm for the MU receiver and −56 dBm for the LP receiver. They show the amount of offset that can be tolerated before the sensitivity increases by 1 dB. Measurement was done with the fixed transmitter frequency reference and variable reference on the receiver side. Looking at the curves, two parts can be distinguished. The first is at relatively small offsets, where there is only a minor degradation in terms of BER. Then, it can be noticed that after offset increases beyond a certain point there is a sharp increase in BER. This is the region where clock recovery fails, and errors occur in sampling the decoded bits, which results in BER around 0.5. Interestingly, the LP receiver shows less degradation at higher offsets. The reason is that the sub-carrier frequency of the demodulated signal doubles after demodulation by the LP receiver. As a consequence, the frequency deviation will be doubled and the FSK demodulator will be able to operate more reliably at higher offsets. Both receivers are still able to perform reliable demodulation at offsets below 8000 ppm. This value is above the reported frequency deviation in state-of-the-art on-chip reference oscillators. Therefore, the presented measurements clearly show that it is feasible to make a fully integrated FM-UWB transceiver, without external resonators for a reference clock.

The main advantage of the proposed MU receiver is the ability to distinguish multiple FM-UWB signals at the receiver input, provided that the

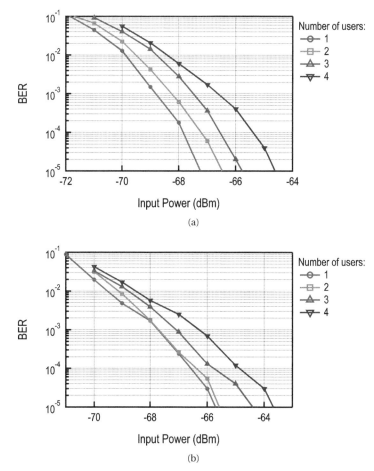

Figure 6.54 Measured BER curves for multiple FM-UWB users of same power level, demodulated with external (a) and internal (b) demodulator.

sub-carrier frequencies are different. In the first scenario, multiple FM-UWB signals of equal power are present at the input. The measurement with the external demodulator is shown in Figure 6.54(a), and the measurement with the internal FSK demodulator is shown in Figure 6.54(b). The curves behave in a similar manner in both cases. As the number of users increases so does the inter-user interference and, as a result, the sensitivity degrades. In the case with 4 users a degradation of roughly 2 dB can be observed from the graph. As long as the power levels at the receiver input remain approximately equal, there will be no significant performance degradation.

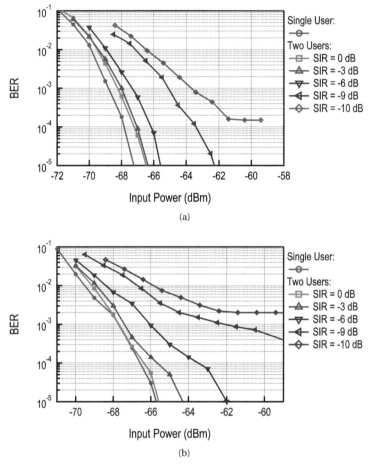

Figure 6.55   Measured BER curves for two FM-UWB users of different power levels, demodulated with external (a) and internal (b) demodulator.

If several transmitting nodes are located at different distances from the receiving node, they can adjust the their power levels so that the power at the receiver remains the same, and avoid the excessive sensitivity loss. This is the main reason why a transmitter that can adjust output power is useful for a HD-WSN.

A scenario with 2 users and a variable difference in power levels is shown in Figure 6.55. Again, the inter-user interference increases with the increase in interferer power, and the resulting sensitivity degrades. With a sufficiently strong interferer at the input, the achievable SNIR after the first

FM demodulation becomes limited by the interference, and the resulting BER curve flattens. The difference between the external and internal demodulation is clearly visible in Figures 6.55(a) and 6.55(b). In both cases 10 dB stronger interferer results in a floor for the achievable BER. Since the external demodulator needs a 1 dB lower SNIR for the same performance, the BER curve flattens below $2 \cdot 10^{-4}$. In the case of the internal demodulator, the curve flattens at $2 \cdot 10^{-3}$, and the desired sensitivity level remains unreachable. With the internal demodulator the receiver is capable of demodulating data if the difference of input levels is 9 dB or smaller. The limit is due to a combination of interference coming from the leakage of adjacent FSK sub-channels and cross modulation products between the two users in the process of the first FM demodulation.

6.6 Summary

A fully-integrated FM-UWB transceiver has been presented in this chapter. A single RF IO port is used, with an on-chip matching network, which eliminates the need for external passive components or switches. Two receivers provide two different modes of operation. The low power mode reduces the power consumption to $267 \, \mu$W, but only allows a single FM-UWB channel. To improve sensitivity and allow SC-FDMA, a MU receiver is used, that allows up to 4 FM-UWB transmitters to operate at the same time and in the same RF band. Inherent robustness against narrowband interferers, combined with sharp IF filtering, result in good in-band and out of band interferer rejection. This capability could allow to power the transceiver wirelessly using a 2.4 GHz narrowband signal. Finally, the transceiver is robust against reference clock offsets of up to 8000 ppm, effectively eliminating the need for an off-chip reference. Performance summary and comparison with the state of the art is given in Table 6.4.

The architecture of the implemented transmitter is very similar to the one from [3]. The solution from [3] uses a three stage DCO that generates one third of the carrier frequency. The three phases are combined using a frequency tripler that generates the FM-UWB signal. The tripler reuses the current of the power amplifier in order to minimize power consumption. One of the downsides of that approach is that the signal at one third of the frequency appears at the output together with its harmonics, which might violate the spectral mask. The second issue is that the tripler output power cannot be precisely controlled, which is a useful feature when a large number of nodes operate in a small area. In the proposed implementation this problem

Table 6.4   Comparison with the state-of-the-art transceivers

Parameter	[27][1,2]	[28]	[3]	This Work	
				LP	MU
Modulation	FM-UWB	Chirp-UWB	FM-UWB	FM-UWB	
Frequency	7.5 GHz	8 GHz	4 GHz	4 GHz	
Frequency deviation	25 kHz	–	250 kHz	50 kHz	
Receiver Cons.	9.1 mW	4/0.6 mW[3]	580 μW	267 μW	550 μW
Data Rate	50 kb/s	1 Mb/s	100 kb/s	100 kb/s	
FSK Sub-channels	2	No	No	No	4
SIR UWB	–	–	–	–	−9 dB
Matching Network	Ext.	Ext.	Ext.	Internal	
NB Interferer Power (in band)	−55 dBm	–	−52 dB[4]	−52 dBm	−55 dBm
NB Interferer Power (@ 2.4 GHz)	−38 dBm @ 6 GHz	–	−38 dB[4]	3 dBm	1 dBm
Ref. Clock Offset	–	–	–	8000 ppm	
Sensitivity	−88 dBm	−76 dBm	−80.5 dBm	−57 dBm	−68 dBm
Transmitter Cons.	–	2.8/0.42 mW[3]	630 μW	583 μW	
Output Power	–	–	−12.8 dBm	−11.4 dBm	
Technology	0.25 μm BiCMOS	65 nm	90 nm	65 nm	

[1] Off-chip sub-carrier FSK demodulation [2] Receiver only [3] Without/with duty-cycling [4] −70 dBm input signal power.

is solved by replacing the trippler with a class AB amplifier that drives the main PA. The output power can be regulated using the amplifier bias point, that also sets the bias of the main PA. In addition, the configurable buffer current, and matching network provide additional knobs for output power control, allowing steps smaller than 3 dB. The DCO directly produces the signal at 4 GHz, avoiding the problem with the spectral mask violation. The DCO together with buffers consumes slightly more than the DCO from [3], but this is compensated with a more efficient PA design, so that the overall transmitter consumption still improves.

The proposed receiver targets short range communication in a HD-WSN. Therefore, sensitivity constraint is not very stringent and emphasis is on reducing power and providing means for multi-user communication. The receiver from [3] achieves very good sensitivity and low power consumption, but the demodulator characteristic is highly non-linear and it cannot support multiple sub-carrier channels. In addition it uses a frequency deviation of 250 kHz (or equivalently 500 kHz separation between the two sub-carrier frequencies), which allows for a simpler FSK demodulator implementation. In this work, frequency deviation is reduced to 50 kHz in order to allow for multiple sub-carrier channels. A different demodulator implementation is necessary in order to demodulate an FSK signal with a modulation index of 1, which will consume slightly more power. Potential for multi-user communication with FM-UWB has been demonstrated before, for example, two different sub-carrier channels could be seen at the demodulator output from [27]. However, this implementation was only providing the first FM-UWB demodulation. The proposed receiver is the only fully integrated solution that provides support for multi-user communication. In addition, it also incorporates a clock recovery circuit that demonstrates the feasibility to integrate the full transceiver with no need for an off-chip crystal reference.

In this case there was no need for higher data rates, although it could be a topic of future research. Increasing data rate of the FM-UWB receiver would mainly require modifications in baseband, and so the overhead in terms of power consumption should remain very low. This should lead to a more efficient implementation, that could achieve even lower energy per bit. Multi-user communication could be explored further, combining either larger number of lower data rate channels, or fewer channels that provide higher data rates, depending on the needs of the specific application. Finally, there could be more room for improvements at the modulation level. The Chirp-UWB concept, that is positioned somewhere between the IR and FM UWB, provides higher data rates, without a significant increase in complexity, and symbol level duty cycling of the receiver provides very low power consumption. Similar modifications could be a topic of future research and could lead to different performance trade-offs.

References

[1] J. Chabloz, D. Ruffieux, and C. Enz, "A low-power programmable dynamic frequency divider," in *Solid-State Circuits Conference, 2008. ESSCIRC 2008. 34th European*, Sep. 2008, pp. 370–373.
[2] P. Nilsson, J. F. M. Gerrits, and J. Yuan, "A low complexity DDS IC for FM-UWB applications," in *2007 16th IST Mobile and Wireless Communications Summit*, July 2007, pp. 1–5.

[3] N. Saputra and J. R. Long, "A fully integrated wideband FM transceiver for low data rate autonomous systems," *IEEE Journal of Solid-State Circuits*, vol. 50, no. 5, pp. 1165–1175, May 2015.

[4] N. Saputra and J. Long, "A Fully-Integrated, Short-Range, Low Data Rate FM-UWB Transmitter in 90 nm CMOS," *IEEE Journal of Solid-State Circuits*, vol. 46, no. 7, pp. 1627–1635, July 2011.

[5] A. Hajimiri, S. Limotyrakis, and T. H. Lee, "Jitter and phase noise in ring oscillators," *IEEE Journal of Solid-State Circuits*, vol. 34, no. 6, pp. 790–804, June 1999.

[6] P. Reynaert and M. Steyaert, *RF Power Amplifiers for Mobile Communications*, ser. Analog Circuits and Signal Processing. Springer Netherlands, 2006.

[7] S. Cripps, *RF Power Amplifiers for Wireless Communications*, ser. Artech House microwave library.   Artech House, 2006.

[8] J. Pandey and B. P. Otis, "A sub-100 mu w mics/ism band transmitter based on injection-locking and frequency multiplication," *IEEE Journal of Solid-State Circuits*, vol. 46, no. 5, pp. 1049–1058, May 2011.

[9] L. E. Franks and I. W. Sandberg, "An alternative approach to the realization of network transfer functions: The N-path filter," *The Bell System Technical Journal*, vol. 39, no. 5, pp. 1321–1350, Sep. 1960.

[10] A. Ghaffari, E. A. M. Klumperink, M. C. M. Soer, and B. Nauta, "Tunable high-q n-path band-pass filters: Modeling and verification," *IEEE Journal of Solid-State Circuits*, vol. 46, no. 5, pp. 998–1010, May 2011.

[11] M. C. M. Soer, E. A. M. Klumperink, P. T. de Boer, F. E. van Vliet, and B. Nauta, "Unified frequency-domain analysis of switched-series-RC passive mixers and samplers," *IEEE Transactions on Circuits and Systems I: Regular Papers*, vol. 57, no. 10, pp. 2618–2631, Oct. 2010.

[12] A. Mirzaei, H. Darabi, and D. Murphy, "A low-power process-scalable super-heterodyne receiver with integrated high-Q filters," *IEEE Journal of Solid-State Circuits*, vol. 46, no. 12, pp. 2920–2932, Dec. 2011.

[13] C. Salazar, A. Cathelin, A. Kaiser, and J. Rabaey, "A 2.4 GHz interferer-resilient wake-up receiver using a dual-IF multi-stage N-path architecture," *IEEE Journal of Solid-State Circuits*, vol. 51, no. 9, pp. 2091–2105, Sep. 2016.

[14] C. Andrews and A. C. Molnar, "A passive mixer-first receiver with digitally controlled and widely tunable RF interface," *IEEE Journal of Solid-State Circuits*, vol. 45, no. 12, pp. 2696–2708, Dec. 2010.

[15] D. Yang, C. Andrews, and A. Molnar, "Optimized design of n-phase passive mixer-first receivers in wideband operation," *IEEE Transactions on Circuits and Systems I: Regular Papers*, vol. 62, no. 11, pp. 2759–2770, Nov. 2015.

[16] M. Darvishi, R. van der Zee, E. A. M. Klumperink, and B. Nauta, "Widely tunable 4th order switched G_m–C band-pass filter based on N-path filters," *IEEE Journal of Solid-State Circuits*, vol. 47, no. 12, pp. 3105–3119, Dec. 2012.

[17] M. Darvishi, R. v. d. Zee, and B. Nauta, "A 0.1-to-1.2 GHz tunable 6th-order N-path channel-select filter with 0.6 dB passband ripple and +7 dBm blocker tolerance," in *2013 IEEE International Solid-State Circuits Conference Digest of Technical Papers*, Feb. 2013, pp. 172–173.

[18] M. Darvishi, R. van der Zee, and B. Nauta, "Design of active N-path filters," *IEEE Journal of Solid-State Circuits*, vol. 48, no. 12, pp. 2962–2976, Dec. 2013.

[19] F. Krummenacher and N. Joehl, "A 4-MHz CMOS continuous-time filter with on-chip automatic tuning," *IEEE Journal of Solid-State Circuits*, vol. 23, no. 3, pp. 750–758, June 1988.

[20] S. D'Amico, M. D. Matteis, and A. Baschirotto, "A 6th-order 100 μA 280 MHz source-follower-based single-loop continuous-time filter," in *2008 IEEE International Solid-State Circuits Conference – Digest of Technical Papers*, Feb. 2008, pp. 72–596.

[21] S. D'Amico, M. Conta, and A. Baschirotto, "A 4.1-mW 10-MHz fourth-order source-follower-based continuous-time filter with 79-dB DR," *IEEE Journal of Solid-State Circuits*, vol. 41, no. 12, pp. 2713–2719, Dec. 2006.

[22] D. Griffith, P. T. Røine, J. Murdock, and R. Smith, "17.8 a 190 nW 33 kHz RC oscillator with ±0.21% temperature stability and 4ppm long-term stability," in *2014 IEEE International Solid-State Circuits Conference Digest of Technical Papers (ISSCC)*, Feb. 2014, pp. 300–301.

[23] K. J. Hsiao, "A 32.4 ppm/°c 3.2-1.6 V self-chopped relaxation oscillator with adaptive supply generation," in *2012 Symposium on VLSI Circuits (VLSIC)*, June 2012, pp. 14–15.

[24] A. Paidimarri, D. Griffith, A. Wang, A. P. Chandrakasan, and G. Burra, "A 120 nW 18.5 kHz RC oscillator with comparator offset cancellation for ±0.25% temperature stability," in *2013 IEEE International Solid-State Circuits Conference Digest of Technical Papers*, Feb. 2013, pp. 184–185.

[25] T. Tokairin, K. Nose, K. Takeda, K. Noguchi, T. Maeda, K. Kawai, and M. Mizuno, "A 280 nW, 100 kHz, 1-cycle start-up time, on-chip CMOS relaxation oscillator employing a feedforward period control scheme," in *2012 Symposium on VLSI Circuits (VLSIC)*, June 2012, pp. 16–17.

[26] J. Masuch and M. Delgado-Restituto, "A 1.1-mW-RX -dBm Sensitivity CMOS Transceiver for Bluetooth Low Energy," *IEEE Transactions on Microwave Theory and Techniques*, vol. 61, no. 4, pp. 1660–1673, Apr. 2013.

[27] Y. Zhao, Y. Dong, J. F. M. Gerrits, G. van Veenendaal, J. Long, and J. Farserotu, "A short range, low data rate, 7.2 GHz-7.7 GHz FM-UWB receiver front-end," *IEEE Journal of Solid-State Circuits*, vol. 44, no. 7, pp. 1872–1882, July 2009.

[28] F. Chen, Y. Li, D. Liu, W. Rhee, J. Kim, D. Kim, and Z. Wang, "A 1 mW 1 Mb/s 7.75-to-8.25 GHz chirp-UWB transceiver with low peak-power transmission and fast synchronization capability," in *2014 IEEE International Solid-State Circuits Conference Digest of Technical Papers (ISSCC)*, Feb. 2014, pp. 162–163.

[29] A. Ghaffari, E. Klumperink, and B. Nauta, "8-Path tunable RF notch filters for blocker suppression," in *2012 IEEE International Solid-State Circuits Conference*, Feb. 2012, pp. 76–78.

[30] V. Kopta and C. Enz, "A 100 kb/s, 4 GHz, 267 μW fully integrated low power FM-UWB transceiver with multiple channels," in *2018 IEEE Custom Integrated Circuits Conference (CICC)*, April 2018.

[31] V. Kopta and C. C. Enz, "A 4-GHz low-power, multi-user approximate zero-IF FM-UWB transceiver for IoT," *IEEE Journal of Solid-State Circuits*, vol. 54, no. 9, pp. 2462–2474, Sep. 2019.

7

Conclusion

The number of connected devices keeps increasing, but to truly realize the vision of IoT, the vision of thousands of wireless nodes per person, cost, size and consumption of such nodes must be lowered. Cost is related to the size of the node and the number of components. Reducing the number of components is always beneficial, and ideally the whole system would be integrated on a single silicon die. With advances in silicon technologies, various electronic components have become so small that the biggest component on a sensor node is in fact the battery. The smaller it is, the smaller the node will be. Unsurprisingly, the size of the battery is related to its capacity and, consequently, the lower the consumption of a node the smaller the battery and its size.

The consumption of a sensor node is commonly dominated by the consumption of its radio, and lowering the consumption of a radio remains the biggest challenge of the IoT. At the same time, as the number of devices continues to grow, other issues will appear: the increasing number of interferers, scalability problems and latency, to name a few. In this work, we propose FM-UWB as a solution to all of these needs. Inherent interference rejection and support for multiple sub-carrier channels appear to be promising to provide good scalability and robust communication in a progressively noisy environment. Simple receiver and transmitter architectures guarantee low peak power consumption allowing to use smaller batteries. Potential to integrate all components of the transceiver has been demonstrated in the preceding chapters, making FM-UWB the ideal choice for a miniature sensor node.

7.1 Summary of Achievements

As discussed previously, when it comes to power consumption, narrowband receivers inevitably have the advantage. The FM-UWB receiver simply cannot achieve the same sensitivity for a given power consumption, but it brings other benefits to the table. The work presented here continues the trend of lowering the power of FM-UWB transceivers, as illustrated in Figure 7.1. The implemented single-ended AZ-IF receiver further narrows the gap between the narrowband and wideband receivers, but it does so at the cost of lower sensitivity. The implemented quadrature AZ-IF receiver consumes more power, but allows up to four FM-UWB transceivers to share the same RF band, and provides better sensitivity.

The trade-off between sensitivity and efficiency, for different FM-UWB receivers is shown in Figure 7.2. A third axis that could be added, and is often neglected, could be linearity, that is closely related to the ability to distinguish multiple FM-UWB signals. The RF delay line demodulator consumes the most, but still provides the best sensitivity performance, and should be able to support multiple sub-carrier channels [1]. The super regenerative architecture [2, 3], achieves considerable power savings, and relatively good sensitivity, but has a very non-linear demodulator characteristic. Finally, the two architectures proposed in this work are targeting short-range communications, and can therefore withstand lower sensitivity in order to further reduce the

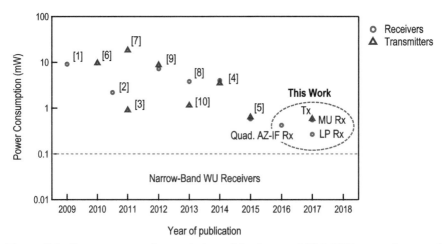

Figure 7.1 Power consumption evolution of implemented FM-UWB transmitters and receivers.

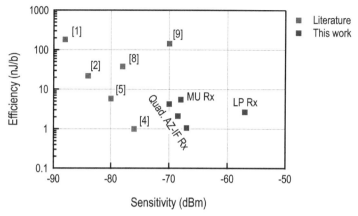

Figure 7.2 Efficiency vs. sensitivity of implemented FM-UWB receivers.

power requirements and provide enough linearity to enable the multi-user communication.

The core of the work focuses on the implementation of an ultra low power FM-UWB transceiver. The design process started with the exploration of options and capabilities of the FM-UWB modulation. Architectures that could yield the desired power reduction have been studied. Finally, different low power circuit techniques have been used to arrive to the achieved power consumption. The main contributions are summarized here:

- The FM-UWB modulation, originally proposed by Gerrits, is a combination of the low modulation index BFSK at baseband, and the large modulation index FM at RF. The concept was first extended to include other baseband modulation types such as MFSK, BPSK and MPSK. Here, a multi-channel transmission, similar to OFDM, is proposed, using a sum of several orthogonal sub-carrier signals to modulate the carrier. The concept allows simultaneous transmission in the same frequency band to multiple receivers and is demonstrated through experiments in Chapter 5.

- The approximate zero-IF architecture is introduced as a modification of the uncertain IF narrowband receiver. The quadrature approximate zero-IF receiver allows to save power while preserving enough FM-AM conversion linearity to allow simultaneous demodulation of multiple input FM-UWB signals. The single-ended version of the receiver further simplifies its architecture reaching the record low power consumption, but loses sensitivity and loses the capability to distinguish multiple FM-UWB users.

- Aside from power consumption, the implemented transceiver is the first full FM-UWB transceiver that provides the multi-user capability. In this case, four channels with a 100 kb/s data rate are available for simultaneous communication. High resilience to out of band interferers enables reliable communication in the presence of other devices, in particular the devices operating in the 2.4 GHz ISM band.

- High tolerance to large reference frequency offsets is demonstrated, and a circuit that compensates for the mismatch between the transmit and receive clock is proposed. This allows for the integration of the FM-UWB full transceiver without the need for an external high-Q resonator, or any other off-chip components.

- Electronic circuits have been a topic of research for a long time, making it rather difficult to innovate and discover truly new topologies. Most of the used circuits are adaptations and minor modifications of already existing solutions. The two more notable blocks that brought some innovation in this work are the baseband N-path channel filter and the low power DCO. The use of an N-path filter in the baseband allows to easily implement a tunable, high-Q bandpass filter, with relatively low power consumption and simple control. Although N-path filters have been out there for some time, they have never been used in a low power and low frequency application such as this one. Implementing the DCO as a stack of a ring oscillator and a frequency multiplier allowed to run the oscillator at one third of the actual carrier frequency, allowing it to reduce consumption. This was one of the key innovations that led to such low overall consumption of the single-ended approximate zero-IF receiver.

7.2 Future of FM-UWB

The aim of this work was to demonstrate that the FM-UWB answers to most of the modern needs of the IoT, and highlight its advantages compared with conventional narrowband radios. The FM-UWB could develop in several directions:

- Miniaturization – as shown here, the FM-UWB transceiver can operate without a precise frequency reference such as a quartz oscillator. Removing yet another component from a sensor node allows to reduce both size and cost. However, most protocols today rely on some form of time keeping. Efficient asynchronous schemes are needed at a higher level in order to truly exploit such a transceiver and achieve the desired degree of miniaturization.

- Multi-user communication and scalability – a case with 4 sub-carrier channels is demonstrated here, but a different number of channels could be used as well. Techniques that would allow for a larger sub-carrier band and provide more channels can be explored (e.g. higher SC frequencies). At the same time, a trade-off between a number of channels and the data rate per channel can be explored to optimize performance for a particular application. An adaptable solution can be envisioned, allowing the radio to adapt to the changing conditions in a WSN.
- Low power – different methods, not studied here, could allow to find a better trade-off between noise and power consumption. For example, a receiver based on a PLL or a frequency tracking loop has not yet been studied in the context of FM-UWB. New architectural and circuit approaches could enable to further lower the power consumption and potentially achieve the levels of narrowband wake-up receivers.

References

[1] Y. Zhao, Y. Dong, J. F. M. Gerrits, G. van Veenendaal, J. Long, and J. Farserotu, "A Short Range, Low Data Rate, 7.2 GHz-7.7 GHz FM-UWB Receiver Front-End," *IEEE Journal of Solid-State Circuits*, vol. 44, no. 7, pp. 1872–1882, July 2009.

[2] N. Saputra and J. Long, "A Short-Range Low Data-Rate Regenerative FM-UWB Receiver," *IEEE Transactions on Microwave Theory and Techniques*, vol. 59, no. 4, pp. 1131–1140, Apr. 2011.

[3] ——, "A Fully-Integrated, Short-Range, Low Data Rate FM-UWB Transmitter in 90 nm CMOS," *IEEE Journal of Solid-State Circuits*, vol. 46, no. 7, pp. 1627–1635, July 2011.

[4] F. Chen, Y. Li, D. Liu, W. Rhee, J. Kim, D. Kim, and Z. Wang, "9.3 A 1 mw 1 mb/s 7.75-to-8.25 ghz chirp-UWB transceiver with low peak-power transmission and fast synchronization capability," in *Solid-State Circuits Conference Digest of Technical Papers (ISSCC), 2014 IEEE International*, Feb. 2014, pp. 162–163.

[5] N. Saputra and J. R. Long, "A Fully Integrated Wideband FM Transceiver for Low Data Rate Autonomous Systems," *IEEE Journal of Solid-State Circuits*, vol. 50, no. 5, pp. 1165–1175, May 2015.

[6] B. Zhou, R. He, J. Qiao, J. Liu, W. Rhee, and Z. Wang, "A low data rate FM-UWB transmitter with-based sub-carrier modulation and quasi-continuous frequency-locked loop," in *2010 IEEE Asian Solid-State Circuits Conference*, Nov. 2010, pp. 1–4.

[7] B. Zhou, H. Lv, M. Wang, J. Liu, W. Rhee, Y. Li, D. Kim, and Z. Wang, "A 1 mb/s 3.2-4.4 ghz reconfigurable FM-UWB transmitter in 0.18 μm CMOS," in *2011 IEEE Radio Frequency Integrated Circuits Symposium (RFIC)*, June 2011, pp. 1–4.

[8] F. Chen, W. Zhang, W. Rhee, J. Kim, D. Kim, and Z. Wang, "A 3.8-mW 3.5-4-GHz Regenerative FM-UWB Receiver With Enhanced Linearity by Utilizing a Wideband LNA and Dual Bandpass Filters," *IEEE Transactions on Microwave Theory and Techniques*, vol. 61, no. 9, pp. 3350–3359, Sep. 2013.

[9] B. Zhou, J. Qiao, R. He, J. Liu, W. Zhang, H. Lv, W. Rhee, Y. Li, and Z. Wang, "A Gated FM-UWB System With Data-Driven Front-End Power Control," *IEEE Transactions on Circuits and Systems I: Regular Papers*, vol. 59, no. 6, pp. 1348–1358, June 2012.

[10] F. Chen, Y. Li, D. Lin, H. Zhuo, W. Rhee, J. Kim, D. Kim, and Z. Wang, "A 1.14 mw 750 kb/s FM-UWB transmitter with 8-FSK subcarrier modulation," in *2013 IEEE Custom Integrated Circuits Conference (CICC)*, Sep. 2013, pp. 1–4.

Index

A

active mixer 69, 84, 140, 174
adjacent channel 14, 50, 147
AM signal 54, 55, 79
approximate zero-IF 85, 89, 125, 191
attenuation 97, 148, 172, 173
average frequency 159
average power 23, 26, 57, 121
AWG 175
AZ-IF 7, 8, 125, 190

B

balanced frequency discriminator 57, 58
baseband signal 14, 175
battery 5, 16, 63, 189
biquadratic section 150
BLE receiver 12, 17, 61, 87
BLE transmitter 13
Bluetooth 1, 18, 84, 165

C

carrier offset 17, 74, 112, 165
Carson's rule 40, 49
center frequency 40, 70, 161, 177
channel filter 92, 127, 139, 192
class AB 60, 135, 136, 184
clock recovery 125, 158, 180, 185
CMOS 2, 68, 155, 162

consumption breakdown 170, 174
current DAC 100, 127, 130
current reference 106, 107
current reuse 16, 60, 68
current steering 130, 131, 139, 163

D

DCO buffer 101, 135, 169, 174
delay path 103
delay-line demodulator 103, 120
demodulator bandwidth 71, 81, 103, 118
die photograph 107, 162
digital sub-carrier 127, 129, 130
direct conversion 14, 16, 90, 98
distortion 42, 74, 148, 153
dual band-pass filter 57
Duty-cycling 57, 121, 184

E

envelope detector 18, 57, 86, 146

F

feedback amplifier 95
filter transfer function 153
FLL calibration 59, 129, 159, 171
FM demodulator 41, 79, 114, 175

FM-AM conversion 55, 83, 121, 191
frequency deviation 40, 110, 180, 185
frequency divider 15, 59, 132, 171
frequency offset 54, 80, 115, 192
frequency selective fading 44, 61
frequency step 100, 171
frequency trippler 144, 171, 183
FSK channel 49, 50
FSK demodulator 43, 110, 161, 185
fully integrated 54, 125, 183, 185

G
gain control 97, 105, 141
Gerrits' approximation 83
Gilbert's mixer 103

H
high-speed 39

I
IF amplifier 69, 91, 141, 179
in-band interferer 116, 178
injection locking 22
input SNR 42, 44, 84
inter-user interference 77, 117, 181
interference rejection 9, 45, 58, 189
Internet of Things 2, 3, 9
IQ delay line demodulator 79, 91
IR-UWB 23, 39, 51, 63
ISM band 13, 21, 179, 192

K
Krummenacher differential pair 150

L
linear PA 60, 135
link budget 12, 25, 28
LO calibration 160
LoRa 9, 12, 191
low noise amplifier 68
low-pass equivalent 150

M
matching network 17, 60, 136, 183
measurement setup 110, 174, 178
modulation index 25, 51, 185, 191
MSO 110, 164, 175
multi-path 11, 48
Multi-user 7, 67, 125, 193

N
N-path filter 21, 148, 173, 192
narrowband 7, 39, 191, 193
narrowband interferer 39, 44, 145, 183
narrowband radio 11, 61, 63, 192
NFC 10
noise floor 45, 77, 115
noise power 43, 47, 48, 84
noise PSD 42
non-coherent 43, 83, 110
non-overlapping 148, 151, 161

O

OFDM 53, 120, 191
optimal demodulator 42
oscillator-trippler 114
out of band interferer 143, 179,
    183, 192
outdoor mask 168
output power 13, 135, 168, 184
output SNR 53, 74, 83
output stage 134, 137, 168, 170

P

PA efficiency 135, 138
parasitic capacitance 94, 104,
    149, 172
PDF 41, 52
peak power 24, 39, 121, 189
performance summary 57,
    61, 183
phase noise 14, 111, 132, 166
power amplifier 14, 134, 168, 183
power control 184
PSG 110, 175

Q

quality factor 49, 98, 148, 153
quartz 11, 20, 22, 192
quasi-stationary 41

R

receiver architecture 20, 39,
    78, 127
receiver sensitivity 82, 84, 176
reference frequency 22, 62,
    159, 192
reflection coefficient 108, 137,
    139, 169
regenerative demodulator 56

RF frontend 95, 111, 139, 142
RF IO 137, 163, 169, 183
RFID 10, 11
ring oscillator 22, 59, 144, 192

S

SC-FDMA 45, 86, 127, 183
sensitivity degradation 46, 112,
    116, 179
sensitivity estimation 82, 86
sensor node 5, 29, 89, 189
sliding IF 14, 15
SNIR 47, 50, 147, 183
spectral mask 41, 52, 168, 184
spectrum analyzer 175, 177
spread spectrum 22, 39, 61, 120
sub-carrier frequency 41, 80,
    128, 180
sub-carrier synthesis 128
switch resistance 149
switch-capacitor 151
switching PA 59, 60, 135
symbol clock 125, 158, 159, 161
synchronization 18, 44, 52, 63

T

test setup 109, 175
transconductor 150, 151, 153
transmitter consumption 58, 184
transmitter DCO 132, 171
two stage ring oscillator 99

U

uncertain IF 21, 68, 86, 191
UWB band 45, 130, 137, 177

V

voltage gain 92, 105, 154, 171

W

wake-up receiver 12, 61, 63, 193
WBAN 51, 177
weak inversion 104, 106, 150, 155
wideband amplifier 70

wideband demodulator 23, 70, 139, 179
WiFi 10, 25
WSN 31, 117, 184, 193

Z

zero crossing 158

About the Authors

Vladimir Kopta received his B.S. degree in electrical and electronics engineering from the University of Belgrade, Serbia, in 2011 and the M.S. and Ph.D. degrees from the Swiss Federal Institute of Technology (EPFL), Switzerland, in 2013 and 2018.

Since 2018 he has been working as an R&D engineer at theIntegrated and Wireless Systems Division of the Swis Center for Electronics and Microtechnology (CSEM), Neuchatel, Switzerland. Focus of his work is the design of low power analog and RF integrated circuits for wireless communications as well as power management for systems on chip.

Christian Enz, PhD, Swiss Federal Institute of Technology (EPFL), 1989. He is currently Professor at EPFL, Director of the Institute of Microengineering and head of the IC Lab. Until April 2013 he was VP at the Swiss Center for Electronics and Microtechnology (CSEM) in Neuchâtel, Switzerland where he was heading the Integrated and Wireless Systems Division. Prior to joining CSEM, he was Principal Senior Engineer at Conexant (formerly Rockwell Semiconductor Systems), Newport Beach, CA, where he was responsible for the modeling and characterization of MOS transistors for RF applications. His technical interests and expertise are in the field of ultralow-power analog and RF IC design, wireless sensor networks and semiconductor device modeling. Together with E. Vittoz and F. Krummenacher he is the developer of the EKV MOS transistor model. He is the author and co-author of more than 250 scientific papers and has contributed to numerous conference presentations and advanced engineering courses. He is an IEEE Fellow and an individual member of the Swiss Academy of Engineering Sciences (SATW). He has been an elected member of the IEEE Solid-State Circuits Society (SSCS) AdCom from 2012 to 2014 and was Chair of the IEEE SSCS Chapter of Switzerland until 2017.